Guide to
HOMEBUILTS

by

Peter M. Bowers

NEW YORK

MODERN AIRCRAFT SERIES

A DIVISION OF SPORTS CAR PRESS

MODERN AIRCRAFT SERIES

Edited by Joe Christy

A new series of popular-priced books on aircraft and their operation for everyone interested in privately owned planes. Each volume will be written by an expert in the field, and will be printed on fine, white paper and be profusely illustrated with photographs and diagrams.

Cockpit Navigation Guide	by Don Downie
Used Plane Buying Guide	by Jim Triggs
Pilot's Weather Guide	by Frank Kingston Smith
Computer Guide	by Joe Christy
Guide to Homebuilts	by Peter Bowers
Cessna Guide	by Lindy Boyes
Light Plane Engine Guide	by Leslie Thomason
Racing Planes Guide	by Roy Wieden
Air Traffic Guide	by Robert Smith
Modern Aerobatics Guide	by Harold Krier
The Piper Cub Story	by Jim Triggs
Classic Biplanes	by Robert Smith
Aviation Radio for Pilots	by Jim Holahan
Agricultural Aviation	by Alan Hoffsommer
Beechcraft Guide	by Joe Christy
Parachuting for Sport	by Jim Greenwood
Antique Plane Guide	by Peter Bowers
Bomber Aircraft Pocketbook	by Roy Cross
Fighter Aircraft Pocketbook	by Roy Cross

LIBRARY OF CONGRESS CATALOG CARD NUMBER: 62-18925

© 1962 by Sports Car Press, Ltd.

Third Printing, October, 1965

Published in New York by Sports Car Press, Ltd., and simultaneously in Toronto, Canada, by Ambassador Books, Ltd.

All rights reserved under International and Pan American Copyright Convention.

Manufactured in the United States of America by Indiantown Printing, Inc.

Distributed by CROWN PUBLISHERS
419 Fourth Avenue, New York, N. Y.

Contents

Chapter	Page
1. History	5
2. Organization and Usage	16
3. Design and Construction Trends	21
4. Sources of Designs and Materials	44
5. Original Design Procedure	52
6. Construction Problems	78
7. Testing	86
8. Legal Problems and Paperwork	95
9. Costs	106
10. Ground Transportation	110
11. The Top Homebuilts	116

1. History

Amateur building of aircraft for recreational purposes began shortly after heavier-than-air flying became a reality. As early as 1910 such "do-it-yourself" magazines as *Popular Mechanics* were publishing plans for simple hang gliders, and the 1912 edition of Hayward's general aviation text, "Practical Flying," presented plans for both the Curtiss biplane and the Bleriot monoplane. However, there was at this time little distinction between the amateurs, the serious experimenters, and some of the would-be manufacturers, so the actual hobby aspect of do-it-yourself aviation can be considered as originating shortly after the end of World War I.

With thousands of wartime trainers and scouts available at low surplus prices, there was little purpose in building an airplane just to get into the air or to save money. The amateur builder's motivation was to have something different, to work out his own ideas, or just to experience the thrill of taking to the air in a machine of his own creation.

The First Big Boost

The movement, carried on by various individuals without benefit of organization, got its first boost in 1924 when two races for planes with useful loads of 150 pounds and powerplants of 80 cubic inches or less were scheduled for the National Air Races held at Dayton, Ohio. Prior to that time, all races had been for the larger factory-built types, either pure racers, surplus military, or stock commercial types.

The powerplant requirements dictated a size far below what the industry was turning out, and meant that the pilots had to design and build their own machines. Recognition of homebuilts as a part of the American aviation scene dates from those Dayton races. While this recognition didn't result in a boom, it started some serious thinking about the development of ultra-light designs, and this eventually resulted in some of the well-known homebuilts of the late 1920's and early 1930's. One of the 1924 Dayton winners, developed by J. A. Roche, a civilian engineer at the Army's McCook Field, was eventually produced as a certificated lightplane—the Aeronca C-2 of 1929 and it's direct successor, the C-3 of 1931-36.

As the cheap war-surplus planes wore out in the mid-1920's and were replaced by more costly new production types, the economic aspect of the homebuilts became a significant factor. For many people, flying had to be cheap if it was to be done at all, and building one's own light airplane was the only way to do it. While there was still no formal organization, there were enough individual pilots making their wants known for a small-scale industry to develop to meet their needs.

Ed Heath of Chicago developed his famous "Parasol" by using the upper wing panels of a surplus WW-I Thomas-Morse Scout and a converted Henderson motorcycle engine. The Lincoln Airplane Company of Lincoln, Nebraska, took over a small single-seat biplane developed by S. S. Swanson and marketed it as the "Lincoln Sport." And B. H. Pietenpol, who is still actively developing homebuilts, produced a series of "Air Campers" of wooden construction powered by converted Model T and Model A Ford automobile engines in the late 1920's and early 1930's.

Heath won his early fame with his low-powered racers of 1926 and 1927, but built his business selling plans, kits, and factory-built Heath Parasols as well as a wide range of parts and equipment to the homebuilders. After plans for the Parasol were widely published in various aviation magazines, the Heath firm refined it to the point where it was placed on the market as a fully type-certificated lightplane. The Lincoln "Sport" survived the demise of the parent company through publication of the plans in *Modern Mechanix* magazine and its annual compendium of plans and data for the amateur builder, the *Flying and Glider Manual*. Started in 1929, this fifty-cent annual saw its peak years of value to the amateur from 1930 to 1934, and so valuable is the fundamental "How-to-do-it" information in them that copies for those years bring up to $20.00 today, when they can be found.

The "Sport" is still with us today as a result of its basic plan form being incorporated into the production Rose "Parakeet" sportplane of 1934. While this was not successful as a factory design because of the depression, it was suitable for home construction. The design rights were obtained in 1948 by the Hannaford Aircraft Company of Glenview, Illinois, and kits and plans were marketed in the 1950's as the Hannaford "Bee."

First of the "Standard" homebuilts—the Heath Parasol, powered with a converted Henderson motorcycle engine and built from magazine plans and factory-produced kits from 1927 well into the 1930's. (Photo by Boardman C. Reed, 1928)

The 28 H.P. Swanson Sport of 1923, produced by the Lincoln Aircraft Corp. of Nebraska as the Lincoln Sport and available to homebuilders through widely-published magazine plans. (Photo courtesy Erwin J. Bulban, 1923)

Allen Rudolph's Pietenpol "Air Camper" of the 1930's restored and modernized to the extent of using steel-tube landing gear and airwheels, but retaining the original converted Model A Ford automobile engine. (Photo by Peter M. Bowers, 1960)

The 1930-vintage Pietenpol, too, is still very much on the scene. Several of those forced into storage by the restrictions of the mid-1930's have been hauled out and refurbished, while others have been built with modern materials and powerplants from the old magazine and manual plans.

The Peak Years

The peak years of the between-wars period of the homebuilt movement were 1929-1933, after the Lindbergh boom had given impetus to all branches of flying. Publicity, the availability of plans and suitable engines for the amateur, and relative freedom from government regulations covering the situation were great incentives. The "cheap flying" aspect was plugged even harder when the sport of gliding was introduced in 1929—with aircraft and pilots imported from Germany—and several dozen new glider designs made their immediate appearance. Plans appeared in the magazines, kits for simple primary types were available for as little as $100, and some finished factory-built gliders sold for less than $500.

Teaching Yourself to Fly

While the lack of regulations encouraged the amateurs, it also sowed the seeds of their eventual extermination. Some who designed, built, and flew their own machines were qualified pilots and mechanics, but the majority were not. They may have had a few hours of "stick time" with a pilot friend, and may have had a rough knowledge of the fundamentals, but they did not have the proficiency to handle the marginal flying machine that was typical of the homebuilt of that era.

During this period, many who knew nothing at all of piloting undertook to teach themselves in their own machines, using the World War I French method of "grass cutting"—dashing across the back pasture at low power and bouncing the ship into the air to get the feel of it. As the pilot's confidence built up, his hops became longer until he was committed to the problem of crossing the fence and finding some spot straight ahead to set it down on, or of turning it around and bringing it home.

While some succeeded by this method, many more failed. Because of the "Low and Slow" nature of the operation, fatalities were few. But the true number of broken machines and dis-

A modified Rose/Hannaford Parakeet, direct development of the Lincoln Sport, built by Robert Fabian. Performance considerably improved by substitution of 85 HP Continental engine for original 40 HP model in 1934 Rose design. (Photo by Peter M. Bowers, 1960)

Some homebuilts flying today are bona-fide antiques. Sorrell Special flown by author Bowers was originally built in 1930 as White Special powered by 25 HP Henderson, became Smith during 1930's when powered with 40 HP Salmson radial, now boasts 65 HP Continental. (Photo by Alden G. Mowery, 1958)

illusioned pilots will never be known because of the backyard nature of the operations, the understandable lack of publicity, and the absence of present-day requirements for accident reports.

More publicity was given to glider mishaps at this time because gliding, with its manpower requirements for transportation and launching, was essentially a club-type operation from the start and thus was more widely talked about in the community. Then too, the pilots undertook more spectacular flights by being launched from cliff tops and hillsides. The manufacturers of kits were partly responsible for many of the resulting accidents because they pushed sales of their products by saying how easy they were to fly. With such assurances, many eager schoolboys tried to fly crudely-built gliders "by the book" without formal instruction or supervision, and succeeded only in spreading glider parts all over the landscape.

Other Ominous Problems

The lack of supervision extended to the design and construction of the aircraft as well. Many were the weird configurations introduced by the non-conformists who maintained that everyone from the Wright brothers on had been going at it wrong. While some good features did emerge from the home workshops, notably Steve Wittman's now-universal spring steel landing gear, the great majority are mercifully forgotten. A more serious problem than incompetent design was the widespread use of scrounged materials that were nowhere near aircraft quality. For some people, just about any old pine board would do for a wing spar, while glue all too often was the dime store variety. Many wooden parts were held together with clinched nails, and turnbuckles for tightening wire rigging were an unheard of luxury to some builders, even at the going price of 15 cents apiece.

This lack of aircraft quality material was compensated for somewhat by the "Hell-for-Stout" policy of builders who were unable to design a part to a known load factor. They made practically every part several times oversize, and indulged in other structural oddities that contributed to safety by making the machines so overweight that they couldn't fly very high. More accidents were caused by an underpowered and overweight plane failing to clear the fence than by structural failure in the air resulting from understrength parts.

Another saving feature was the fact that a successful flight in most of the early homebuilts was an end in itself, and the pilots of rattletraps didn't care to risk their machines or their necks in aerobatics. Some of the really well-designed types, however, could easily be handled like baby pursuits. Since the gliders were built as light as possible, structural weakness was second only to pilot incompetence as a cause of accidents; engine trouble was the number one headache for the airplanes.

An Appalling Record

Regardless of the specific causes, the safety record of the homebuilts, powered and unpowered alike, was appalling. While increasing regulation was inevitable, it was hastened by the maverick nature of the overall operation. The publicity resulting from accidents, contrasting to the silence about the successful operations, also had an effect on the public and on general aviation to the point where the words "Homebuilt Airplane" became a derogatory term equivalent to "jalopy" for a rickety car. Unfortunately, this connotation still lingers in the minds of many people, in spite of today's radically improved standards and safety records.

The situation wasn't improved any by the introduction of the little French "Flying Flea," which triggered a mild boom in homebuilding in 1936-1937 in those areas where it was still legal. Serious engineering deficiencies in the early models resulted in several fatal crashes and gave homebuilts and amateur designs an even worse reputation than they already had.

Beginning of the End

The Government, through the Bureau of Air Commerce—now the Federal Aviation Agency and hereafter referred to as the FAA—had adopted licensing and airworthiness requirements for interstate flying as early as 1927, but the individual states were allowed to regulate flying within their own borders. One by one, however, they adopted the Federal regulations, which set impossibly high standards on details of design, structural stress analysis, 100% use of aircraft quality materials, etc., as a requirement for the licensing of a new airplane design. Only well-financed companies could meet the costs involved, and the amateurs saw the beginning of the end.

While experimental certificates were issued for non-standard aircraft, they were for such specific purposes as racing, crop dusting, bona-fide development and test work leading to certification, etc. No provision existed for recreational flying in amateur-built aircraft. By the time WW II started, only Oregon still permitted a degree of freedom to the amateur builder-flyer, and a number of them actually moved there from other states just to be able to pursue their hobby.

The other rugged individualists who still wanted to fly homebuilts were forced into bootleg operations in remote areas, so for practical purposes, homebuilt aviation on the national level was dead during the immediate prewar years. The only exceptions were a few racers and exhibition types that were licensed as such and flew only at the occasional races and air shows held throughout the country. The same situation prevailed at the end of WW II, and there was no prospect of immediate change. If anything, the situation was worse because Oregon was now in the Federal fold and the first post-war crop of racers was made up entirely of surplus military types.

Two immensely important events took place in the first postwar years, however, that were to break the bureaucratic barrier and open the way to legal status for amateur-built recreational aircraft.

The Barrier Falls

The first was the 1946 announcement by the Aviation Division of the Goodyear Tire and Rubber Company, that it would sponsor an event at the 1947 National Air Races for a new class of "Baby" racers powered with engines of less than 200 cubic inches, and would support it for three years. Since the new Continental C-85 was the top engine in this category, no baby racers existed that were designed for it. The established industry wasn't interested in producing any, so the announcement of the new racing class was a clear invitation to the amateurs to get active again through an authorized outlet.

Even though there was no legal recognition of the recreational aspect of aviation, and the machines had to be licensed as racers and could only be used as such, this was a start. The pilots had to turn "pro" to do it, however, becoming members of the Professional Race Pilot's Association (P.R.P.A.) in order to fly in the sanctioned races. But the general background remained mostly amateur, fortunately with the sloppy construction and unqualified piloting of the early 1930's safely eliminated by the P.R.P.A. requirements for aircraft airworthiness, performance and pilot qualifications.

Since only one race a year was a small outlet for what was expected to become a national movement, more midget racing

Structural and aerodynamic simplicity of radical French Pou-du-Ciel (Flying Flea) of 1935 gave great promise that it would be the ideal homebuilt design. Serious engineering deficiencies in early models resulted in many accidents and greatly harmed the cause of the homebuilts. (Photo by Myron C. Buswell, 1939)

A representative gathering of Goodyear-class midget racers at Cleveland National Air Races. Construction ranges from all-wood to all-metal and various degrees of combination. Some, like #14 "Little Gem" in foreground, used cut-down wings and tail from production lightplanes. (Photo by Peter M. Bowers, 1949)

was scheduled at other annual events, notably the All-American Air Maneuvers held each winter in Miami, Florida, for which Continental Motors sponsored the midgets. Other races were set up at various local air shows, under P.R.P.A. sanction, with a minimum hazard, a maximum of pilot and spectator enjoyment.

A dozen planes showed up at Cleveland for the 1947 Goodyear Race, the majority of them showing evidence of great haste in construction in order to make this first race. The winner, oddly enough, was Steve Wittman's old 1933 limited-class racer "Buster," which had undergone several engine changes and major modifications in pre-war years and was now completely revamped to take the new engine. That this victory was no fluke is shown by the fact that the same pilot and plane also won the 1949 event against a field of much newer machines.

An Epic Flight

An even more significant event was the round-trip transcontinental flight made by George Beaugardus, one of the keystones of Oregon's prewar activity, from Portland to Washington, D. C., between August and October, 1947. This trip was made for the specific purpose of demonstrating the dependability of amateur-built aircraft powered with standard airplane engines, and to petition the Government for a change in the prevailing regulations. These were still industry-oriented and recognized no recreational aspect of flying.

The plane, a prewar design built by Tom Story of Beaverton, Oregon, and based on the famous Les Long design known as "Wimpy," was modified by George in 1946 with the requirements of the trip in mind. Named "Little Gee Bee" after George's initials, it was fitted with a 65 h.p. Continental, brakes, and other refinements that added to its performance and reliability. It flew under a 90-day experimental license that was issued to it as a demonstration machine. The double coast-to-coast flight was highly successful in that the officials were impressed by the safety record and the petition was not only received and considered, but acted upon. In 1948 a new "Amateur-Built" aircraft category was established which for the first time gave full official status to amateur builders and flyers, and set up logical regulations under which they could operate.

Probably the most significant single home-built airplane ever produced in the United States—the Bogardus "Little Gee Bee" that was flown from Portland, Oregon to Washington, D. C. in 1947 to petition the FAA for recognition of amateur aviation. (Photo from A. U. Schmidt Collection)

Death of the Races

Since the pilots of the midget racers were "pro's" because of the prize money involved and their P.R.P.A. affiliations, and their ships carried experimental licenses issued specifically for racing, they were not affected by the new category. However, the new outlet was far more appealing than the rough competition of racing to most builder-pilots, and interest in the racers waned until only a few dedicated zealots were left.

Goodyear dropped its sponsorship after the third year, and while Continental Motors took over the event, it was dropped entirely after 1951 and the National Air "Races" became the National Air "Show," without competitive events. The local races all but died out, too, partly from loss of builder-pilots to the amateur activities but largely to the difficulty of holding a high-visibility event like a race where free-loading spectators couldn't see almost as much from outside the fence as from the revenue-producing grandstand.

It took a while for the full realization of the fact that home-builts were now legal to sink in, and it was not until the early 1950's that the present boom began to roll.

2. Organization and Usage

Home-building of airplanes began as an individual activity and remained that way for the years between the wars. This was one of its greatest weaknesses, and interfered seriously with any growth of the activity that could normally have been expected. As government regulation and subsequent restrictions built up in the 1930's, a few scattered attempts were made to organize the amateurs for their own defense and survival, but by then it was too late.

A second major handicap of lone wolf operations was the general lack of technical data exchange and experience retention. A builder might get what he thought was a good idea and go to a lot of trouble to try it out, not realizing that a dozen others had already done the same thing and found that it didn't work. There was a certain amount of publicity given to the homebuilts in the pages of the mechanical magazines, and in one general aviation publication, *Popular Aviation*. But these were mostly short "What the Readers are Building" items, with one photograph, and were not serious forums for data exchange. Successes were extolled and failures ignored.

The "Flying Club" concept had little effect in the homebuilt field from its inception. While the desirability of extra hands to share the work and expenses was obvious, amateur aircraft design remained largely an individual project. Planes were often built in partnership, but these were usually designed entirely by one of the partners or plans were procured for an established design. The glider pilots, largely because of the manpower requirements of their operations, were the first to introduce club-type organization in the field of amateur-built aircraft, and were the first to organize on a national scale — the Soaring Society of America being founded in 1932. Powerplane flying clubs were and still are mostly cooperatives built around group ownership of one or more planes for the purpose of reducing the per-hour costs to the members.

Comes The Organization Man

Effective attempts at national organization in the amateur airplane field did not begin until after WW II. The first—the American Airman's Association—was formed by George Beaugardus in

Home-building of aircraft is not exclusively a man's hobby. Not only does Joan Trefethen fly this modified Stits "Playboy," she built it all herself, even to the welding. (Photo by Peter M. Bowers, 1960)

Oregon in 1946. This one suffered from the fact that the new freedom had not yet been established and there was relatively little support even from the zealots because war surplus trainers and liaison planes were available—at less than the cost of materials for a new home-built—for those who wanted to fly.

The timing was better when the Experimental Aircraft Association (EAA, 9711 W. Forest Park Drive, Hales Corners, Wisconsin) was formed in January, 1953. The amateur flyers were alert and ready this time. Cheap flying in production aircraft was getting harder to come by. The surplus pool was exhausted and the flying schools and rental operators had almost completely replaced their pre-war and early post-war 65-h.p. equipment with newer and much more expensive 85-h.p. models. And these were virtually deluxe winged automobiles rather than the older bare-minimum but sportive flying machines. With history thus repeating itself, home-built aircraft again offered very significant economic advantages to the recreational pilot.

Educational aspects of EAA chapter meetings are demonstrated as Stan Weeks of Chapter 26 (Seattle) describes construction of metal wing ribs that he built. Note chart illustrating development of tapered airfoils on wall. (Photo by Peter M. Bowers, 1961)

There were other helpful changes in attitude, too. In addition to a now-tolerant Government, do-it-yourself in general had won wide social acceptance in the postwar years; no longer was a backyard mechanic regarded as the neighborhood nut. With less opposition on the neighborhood and family level, the growth of EAA became phenomenal. This was closely followed by the formation of a related but separate organization, the Antique Airplane Association (AAA, P. O. Box 14, Fremont, Iowa) that directed its do-it-yourself energies to the restoration of old airplanes rather than the creation of new ones. (Actually, some of the antiques are so extensively rebuilt that they are licensed as home-built replicas rather than restored originals and their organized activities are similar to those of EAA.)

While the new organizations were national in scope from their inception, operating through their publications mailed from central offices, local chapters were soon established as individuals in various areas became aware of each other. Now the boom was really underway.

The function of these organizations at the chapter level is principally that of a mutual admiration and encouragement society and a talent pool. Many people have long wanted to build an airplane, either of their own design or from purchased plans, but with no one else around to encourage them or offer help, they hesitate to take the first step. The chapter comes to the rescue by providing plenty of examples of "it can be done," and the variety of experiences and skills of other members can be applied to problem areas. A good bull session held over a new design while it is still on paper can be beneficial to the members at large as well as to the designer.

On the national level these same organizations, through their publications, disseminate technical information that the individuals and local units would have difficulty obtaining on their own. They serve as an experience retention pool through reports received from members concerning their experiments and, most important of all, form a single voice representing thousands of individuals in dealing with the Government in regulatory matters, obtaining special discounts from vendors, and organizing national-level activities that give the movements a sense of unity and purpose. How well this has succeeded in recent years is

A portion of the 135 homebuilts at the 1962 EAA National Fly-In at Rockford, Illinois. Good photos of individual machines are hard to get because they are surrounded by swarms of interested spectators. (Photo by Peter M. Bowers, 1962)

shown by the fact that the Government now consults the organizations when considering changes in the regulations!

In organization and operation, the chapters are more like sports car clubs or small yacht clubs than the standard airplane or glider clubs. (For some unknown reason, clubs operating airplanes are known as "Flying Clubs" while those operating gliders are "Glider Clubs"). While a few chapters do have projects for group building of a single airplane, the majority of them are made up of individuals who own or are building their own machines.

The Fly Ins

Much of the flying activity for homebuilt airplanes originates at the chapter level. "Fly ins," equivalent to sports car rallies, are held frequently and provide highly successful group activity. These are usually invitational, bringing in the members of a nearby chapter and unaffiliated individuals for a day or weekend of flying, eating, showing movies of earlier activities, technical discussions, and general social activity. The flying events are

The ultimate in recreational flying—going somewhere or just flying locally in a pair of evenly-matched homebuilts. The planes are 65 HP Story "Specials" with welded steel tube fuselages and wire-braced wooden wings. (Photo by Kenneth M. Hovik, 1958)

competitive — spot landing, flour bombing, and balloon-busting, for example. Close-course and point-to-point racing is not undertaken. Aside from the competitive events, the pilots spent a great amount of time just studying each other's machines and even trading flights; this is the operation of a mutual admiration society at its best.

As the "Fly-in" activity moves up from the local to the regional level, it grows in size. Greater attendance incentive is provided by additional prize categories — oldest airplane or oldest pilot, airplane flying the greatest distance to attend, "Best" airplane in a particular class (which can stir up all sorts of disagreement), chapter with most members in attendance, etc. Well-publicized regional meets often draw more than 50 homebuilts, plus a scattering of antiques, while the 1962 National EAA Fly-in had 135 drawn from all parts of the country with nearly double that number in "standard" types flown in by pilots who just came to see the show.

Your Greatest Satisfaction

Activity is not limited to such organized group effort, however. Most of the hours are built up by the individual pilot who flies when and where he wants to for proficiency, personal transportation, aerobatics, or just the joy of it all. Straight-away formation flying with another plane of comparable performance, which can be done by practically any pilot, is a very common and enjoyable activity. Dogfighting is resorted to only by skillful pilots with rugged aerobatic machines. One of the greatest satisfactions that can come to the individual, and one of the things that makes all the work worth while, is the attention that he gets when he brings a well-made and attractive homebuilt into an airport where only the stock winged automobiles are based. The crowd gathers, people poke and prod, and then the questions begin. Over a period of time, you'll find that they generally come in this order: "Did you build it yourself, mister?" or "What is it?," meaning what make and model of plane, not necessarily "Is it an airplane?"

"How fast does it go?" "How much did it cost?" The satisfaction that goes with being able to rattle off authoritative answers while in the center of an admiring throng is just as much a part of owning your own homebuilt as roaring off into the wild blue yonder.

3. Design and Construction Trends

One of the notable characteristics of the typical homebuilt airplane is its extreme conservatism in configuration, structure and equipment. In fact, many are so reduced to the bare essentials that their builders are accused of producing latter-day antiques. Since these airplanes are in effect pieces of non-competitive sporting equipment, mere speed — which for a given horsepower is usually obtained at the cost of structural refinement or a decrease in other desirable characteristics — is not a primary design goal. This is understandable when it is realized that the majority are built to provide pleasant and trouble-free recreational flying for people who, in the majority, are essentially "Sunday pilots."

CONFIGURATION

There are unconventional designs and structures, of course, but they form a distinct minority. Throughout nearly 60 years of powered flight, just about every conceivable arrangement of major aircraft components has been tried by somebody. The configurations that prevail today have remained essentially unchanged since 1909 and have proved their overall reliability in the face of competition from other arrangements which, while they may have demonstrated certain special advantages, have usually proved to have a greater number of off-setting disadvantages.

Today's sportplanes are almost universally of the "Tractor" type, with the powerplant in the nose. Actually, the "propeller" is incorrectly named when used in this location; the British are correct in calling the tractor prop an airscrew. The wing is right behind the engine, which "pulls" the plane forward. A few "pusher" designs, with the powerplant reversed to locate the now properly-named propeller behind the wing, or even the tail, appear from time to time but are usually special purpose vehicles. Even then, the traditional wing-in-front configuration is retained.

The home builders found out in short order that the efficiency of biplanes fell off drastically as size decreased, so the majority of homebuilts are monoplanes in three recognized classes:

> *Parasol,* with the wing raised entirely above the fuselage. This can be regarded, because of the necessary supporting struts, as a biplane without a lower wing.

High Wing, in which the wing is flat on top of the fuselage, most generally on cabin types.

Low Wing, with the wing below or flush with the bottom of the fuselage and sometimes supporting the landing gear.

These structures can be strut, wire-braced or full cantilever, and each has its advantages. In the case of homebuilts, these are usually a matter of pilot preference rather than technical logic, although the low wing does offer certain structural and pilot-access advantages, and gives unquestioned superiority in pilot visibility.

In keeping with the sporting character of the operation, most homebuilts are open cockpit types where the pilot can exult in "feeling the wind." Quite a few of these incorporate sliding canopies that make cold weather flying endurable. In spite of their lesser overall efficiency, biplanes do have certain advantages that appeal to some pilots and comprise roughly 20% of the total. Such unconventional configurations as Delta wing (flying dart), Canard (tail surface in front) tandem-wings, and flying automobiles do exist, but these are almost entirely the efforts of a very small experiment-minded minority.

Rotorplanes: A Recent Phenomenon

A relatively recent phenomenon is the popularity of rotorplanes, which are presently encountered in three forms:

Gyro-Glider, an unpowered whirlybird that is kept in the air only when being towed at the end of a rope by an automobile. As long as the ship remains "on the string" the FAA regards it as a kite, and is not concerned with airworthiness certificates, aircraft registration, or pilot certificates. If the bird is cut loose it becomes a glider and is subject to all the prevailing regulations.

Autogyro, widely marketed in kit or plan form under the somewhat misleading name of Gyrocopter. This is essentially the gyro-glider fitted with a powerplant and propeller. A common feature of both is the unconventional control system used in the standard versions. Instead of the traditional stick moved to the left for left bank, etc., the system is reversed by having the stick project downward from the movable rotor head so that left turn is initiated by a move to the right. This presents no problem to the non-pilot who learns on one "by

A representative homebuilt—the Solvay-Stark "Skyhopper," a low-wing single-seat monoplane designed in 1946, with plans still available. Good performance on 65 HP achieved by wings with span slightly longer than average. (Photo by Peter M. Bowers, 1960)

A relatively rare homebuilt type—the all-metal four-place Wickham Model "A" with 135 HP. Costs increase rapidly as airplane size increases, and the final product is practically in direct competition with commercial products. (Photo by Peter M. Bowers, 1957)

Prop-behind-the-tail pushers like the "Nomad" SN-1 are extremely rare in both homebuilt and commercial aircraft and, while retaining conventional "Tractor" aerodynamic layout, are handicapped by mechnical complexity. (Photo by Peter M. Bowers, 1962)

the book," but is a source of great annoyance to pilots with long-established habits of conventional control. They frequently redesign the machines to use standard controls.

Helicopter, which has its own specialized control system consisting of cyclic and collective pitch control, torque-compensating tail rotor or contra-rotating main rotor, and greatly increased mechanical and operational complexity.

The initial popular appeal of these last two classes as "go anywhere" machines is high, but they have not yet been perfected to the point where they are practical transportation. Most are little more than flying milk stools, with precious little lift margin for fuel or baggage and no airstream protection for the one-man crew. These factors don't bother the gadgeteers, but are a serious drawback to the pilot who is looking for utility.

An entirely new form of aircraft now attracting the attention of the gadgeteer is the *Ground Effect Machine (GEM)*, which is not a true flying machine at all but is held a few inches off the ground by the airstream from a blower or propeller leaking out around the base of a plenum chamber that forms most of the bulk of the vehicle. While these are not yet "Transportation," they offer fascinating problems in getting off the ground with low horsepower and in developing suitable directional control.

The size of homebuilt airplanes, most of which are single-seaters for both cost and size-of-the-job considerations, stays in a fairly constant range. And with good reason. Aerodynamic efficiency falls off in the small sizes and the littlest ones are very tricky to fly and, most important, to land. The P.R.P.A. took early note of this and set a minimum of 63 square feet on wing area to keep the landing speeds of the Postwar-II midget racers within reasonable limits. Larger sizes have the disadvantage of being bigger and costlier for the builder, and enter the area already well served by established industry, thereby defeating one of the main purposes of homebuilding — to turn out something that is not already available in the open market.

Small and Practical

The smallest practical designs, therefore (most of them being biplanes in order to retain adequate wing area), are in the 16-20 foot wingspan range with powerplants starting at 65 h.p. Most

While it took the jet engine to make a practical military airplane of the delta wing configuration, a few dedicated amateurs like Marion Baker of Dayton, Ohio, have produced successful propeller-driven types such as Baker's "Delta Kitten." (Photo by Peter M. Bowers, 1962)

monoplanes used for other than racing/exhibition purposes — even the little Volkswagen-powered European types — are found in the 20 to 25-foot range. A few monoplanes designed primarily for easy flying rather than zip sacrifice high-speed performance by going to 28 feet. Empty weights — that is, everything but pilot, gas, and baggage — seldom runs below 500 pounds, except for the V-W jobs, and 550 to 750 pounds is the average for most single-seaters. Gross weights are in the 800-1,000 pound range for single-seaters of 65 h.p. and up and 1,100 to 1,400 for two-seaters to 125 h.p.

Landing Gear

The landing gear configuration chosen for the majority of home-builts is the old-fashioned type referred to as "Conventional." That is, with the main landing wheels forward, where they are ahead of the center of gravity of the airplane. and a tail wheel at the extreme rear of the fuselage. This configuration is more compatible with the type of operation for which most homebuilts are designed. The size and construction of many amateur designs is such that they can use a complete standard landing gear from a small production airplane like a Piper "Cub" or Aeronca "Champion."

Tricycle landing gear, almost universally adopted by the industry since 1950 but actually a pre-WW-I configuration abandoned during hostilities when military performance requirements suppressed mere "safety" features, is not too widely accepted by the homebuilers. Even when it is adopted it is used mostly on cabin models and two-seaters that are more closely related to production types than to sporting models. In small planes, the extra weight of the third wheel and its supporting strut can be a handicap. Then, too, the handling characteristics of the little ships are such that the inherent easy landing and ground steering features of the tricycle gear, so essential to the sale of winged automobiles, are of little significance to the pilot of the typical sporting homebuilt.

Very few homebuilts end up as pontoon-type seaplanes. For most small ships, the extra weight is a formidable handicap. Ready-made floats of the right size are not available, and building a set with the necessary strength-weight ratio, plus the rigging, is no small task. Most of the relatively few homebuilt hydros are of the flying boat type, where the hull/fuselage provides the buoyancy. Standard American pontoon seaplanes are exclusively the inherently stable twin-float type, while the single-hull boats are stabilized on the water by small outrigger floats under the wingtips.

Rise of the Replica

A relatively recent phenomenon in the homebuilt movement is the replica airplane. Antiques have been immensely popular, but not everyone who wants one can find a genuine World War I vintage machine to restore. Actually, replicas have many advantages over originals to all but the most ardent purists. They can operate under the relative freedom allowed amateur-built machines (see Chapter 8) while the originals are usually stuck with highly-restrictive straight experimental certificates.

Some actual antiques end up in the amateur-built category because they incorporate so much new material that they can be considered as built from scratch, using only a few original metal parts. Further, a replica can take advantage of later materials and construction techniques in the interest of reliability and safety. It is quite common practice to make a replica fuselage of welded steel tubing instead of using the complex wood-and-metal-fittings trusswork of the original. Other replicas are just as pure as it is possible to make them, and are frequently built right from original production drawings.

A special class of replica is the "midget." In this plane, an old design is scaled down to three-quarter or even half-size to put it in the practical size, weight, and cost range of the average home-built sport plane. The builders of such types are usually willing to accept necessary deviations from true scale to accommodate modern flat-four engines; not to mention an enlarged cockpit that will allow a standard-size man to fit into a scaled-down airplane. While relatively accurate, full-scale replicas are accepted as a normal part of the antique airplane movement, the midgets do not fit in with the antiques and are regarded by all concerned as homebuilts that are merely patterned after the old timers.

STRUCTURE

The structural details of the homebuilts have followed right along with the standards of the industry, although they do frequently take long steps backward in the interest of simplicity. Wood-frame construction with fabric covering was the standard of the 1920's, with obvious short-cuts in the matter of material quality and structural refinement for reasons of economy. Steel tubing for fuselage construction entered the homebuilt picture with the publication of plans for the Heath Parasol in 1927.

While this was normally a welded structure, a version was available for non-welders in which the ends of the supporting members were hammered flat and joined by steel straps and small bolts or, in some cases, with hammered-over nails! By the early 1930's, welded steel tube fuselages, usually with welded steel tail surfaces but wooden wings, all fabric covered, were the accepted standard. The covering was more often bleached muslin than aircraft grade cotton or linen.

Steel tubing has remained the standard in the years after WW II, but the percentage has been dropping recently because of frequent increases in the price of aircraft-grade tubing. All-sheet-metal construction, either with stressed metal skin or fabric on the wings, is almost entirely a postwar-II phenomenon. It is practically universal in today's lightplane industry, but has some drawbacks for the homebuilder. The main stumbling blocks are tooling investment and the need to virtually build the same airplane twice—once for the form blocks and once for actual assembly.

Since the cost of this construction is now competitive with steel tubing, its popularity is increasing as more effort is applied to developing designs and fabrication procedures particularly suited to it. The sheets are joined to each other and their supporting structure by hand-driven rivets, and the various major components are bolted together. Spot welding and metal bonding have not yet appeared in the amateur field.

The forming of compound curves is still the major drawback to metal construction by the amateur. Without the production tooling and power equipment of the established factory, it is all tedious hand work for the builder of a one-only design. Wing ribs and fuselage formers are most commonly formed by cutting out flat aluminum sheet and hammering it to shape over handmade hardwood form blocks. For straight wings where most of the ribs are the same size, construction time can sometimes be saved by taking the blocks and the blanks to an industrial shop that has a hydropress.

Whenever possible, metal designs are worked out so that areas covered by sheet metal are flat or have only a single curve that a flat sheet can be fitted to. Compound curves have to be formed by pressing or bumping. If the area is non-stressed, the skin can be built up of fiberglass over a mould. "Plastics,"—to use the term broadly and cover epoxies, fiberglass, and foams—are finding increasing use for primary and secondary structures in production aircraft, but cost and the exacting bonding procedures rule them out for anything but cowlings and fairings for the homebuilts.

Building with Wood

Wood is still very popular, especially for wings fitted to fuselages of other construction. Because of their basic structural arrangement, wing and tail structures lend themselves easily to all-wood construction with wood-to-wood (glued) joints. The traditional four-longeron square fuselage is highly unsuited to this, since weight and load considerations on fabric-covered designs dictate

Although not usually licensed as such, many pure experimental designs built to develop and test new ideas, such as this "Planemobile," are actually amateur-built by industry standards. Rear wheel drives this flying automobile while on the ground. (Photo from A. U. Schmidt Collection)

joining the longerons, cross-members, and diagonal bracing through complex metal fittings.

It was difficulty in the fuselage assembly area that prompted the industry's switch to welded steel tubing in the 1920's. The only commercial wooden fuselages built since that time have been semi-monocoque types of laminated wood veneer as used on the Lockheeds of 1928-31 or plywood-over-longerons as used by Culver just before and after WW II and the early post-war Mooneys. Only the homebuilts have used four-longeron wooden truss construction since the late 1920's, and even then they substituted wide wood gussets for the metal fittings.

A pair of ubiquitous Bensen "gyrocopters" in flight, the one on the left fitted with airplane-type "Stick" control while that on the right, flown by Igor Bensen, used his standard overhead control. (Photo by Peter M. Bowers, 1960)

The procedures followed in wood aircraft construction are completely unlike those used in cabinet making or boat yards. There are no mortised joints, tennons, etc. All direct wood joining is by gluing, aided by load-distributing corner blocks and plywood gussets. Joints in plywood are made with 12:1 scarfed splices, a very meticulous job and a major deterrent to plywood construction in stressed areas by those who don't feel that they have the skill.

Nails are NEVER used to hold things together permanently. Brass nails, or steel nails suitably cement-coated or otherwise corrosion-proofed, are used to hold gussets and other small members together while the glue dries, and then become totally redundant. In open areas where they can be applied, wire staples are sometimes substituted for small nails. Larger pieces to be glued are held in contact with screw clamps. Major components are joined, and accessories attached, by bolting; this is usually done in areas reinforced with plywood doublers and with the major load-carrying bolt holes lined with metal or fiber bushings.

All-wood designs, with certain changes from the standard between-wars fuselage construction, are now enjoying a lively boom. When the industry switched to all-metal construction after WW II, the manufacture of aircraft plywood was almost completely ended in the United States, and the price skyrocketed to the point where it was prohibitive to the amateur. The soaring fraternity, with war surplus plywood-winged gliders to maintain, really suffered.

This impasse was resolved in two ways; enterprising individuals began to import aircraft-grade birch plywood from Finland at practically prewar U.S. prices delivered, and the FAA permitted the use of high grade marine plywood in homebuilts. The use of marine was not much of a help at first because it was not available in thicknesses under one-eighth of an inch. This imposed a weight penalty when used as a direct substitute on an established design that used lighter gauges, but now there are new designs developed specifically to use it.

Wooden fuselage construction now in use falls into three general types. Those with rectangular cross-section and four longerons connected by wood uprights and diagonals use wood gussets exclusively as corner members, thereby doing away with the troublesome metal fittings and wire bracing of pioneer and early

Not an airplane, but just as intriguing to its builder—Dr. Bertelsen's "Aeromobile," a ground-effect machine powered by a target drone engine. Shadow shows that it is actually off the ground, with directional control provided by movable vanes. (Photo by Peter M. Bowers, 1960)

postwar-I years. The bumps of the gussets are smoothed out for fabric covering by means of lengthwise stringers that hold the fabric away from the basic structure. By proper use of stringers and supporting formers, a rectangular fuselage structure can be rounded out to an elliptical or fully circular cross-section.

A simplified variation of the rectangular fuselage that elim-

The smallest practical biplane, a 12-footer built by Ray Hegy of Marfa, Texas. Efficiency is increased by straight-chord wings to get maximum area and by raising gap-chord ratio considerably above that of standard proportions. (Photo by Peter M. Bowers, 1960)

inates much of the gusseting covers each side of the frame with plywood, which then acts as a large shear web to take flying and ground loads. The longerons and connecting members can be made with weight-saving smaller cross-sections. A plywood fuselage with skin on all four sides is heavier and not appreciably stronger than one with only two plywood sides joined with a truss.

Geodetic Construction

The technique of wrapping strips of wood veneer over a mould and building it up to several thicknesses, made famous by the German Pfalz fighters of WW I and the later American Lockheeds, is not suited to homebuilts. The nearest approach today is "Geodetic" construction, erroneously called "Basket Weave," where thin strips of wood are wrapped spirally around oval formers to form a rigid glue-jointed structure. As with the rectangular-truss types, longitudinal stringers over the diagonals provide smooth support for the fabric covering.

This type of construction is quite controversial among the more analytical builders. Since it is extremely difficult to stress-analyze, the choice of material size, is usually by educated guess, copied from someone else's existing design, or is a combination of the two beefed up with a good Fudge Factor to take care of the unforeseen eventualities. Such a policy largely undoes one of the major advantages of the structure, its relative lightness. Its extreme rigidity opens it to another technical criticism, poor shock resistance, since it cannot "give" under sudden loads. Damage to the glue joints under such circumstances is hard to detect because of the great number of the joints and the difficulty involved in inspecting them all.

Wooden wing spars and rib cap strips are usually of aircraft grade spruce, although "boat" spruce or good ladder stock is sometimes used. While certain grades of pine, Port Orford Cedar, and other woods are acceptable substitutes as outlined in FAA Manual 18, the saving in cost is hardly worth the extra effort involved in sorting out a flaw-free piece of the desired length. The peace of mind that goes with certified aircraft grade spruce spars is well worth the additional cost. Spars are no items on which to try and save money.

Electrical Bonding: A Peculiar Problem

A peculiar problem of wooden aircraft, never of concern to the amateur until the use of two-way radio became mandatory at all airports served by FAA-operated control towers, is electrical bonding. Isolated metal parts build up varying degrees of static electricity that cause noise in the radio. Industry licked the problem for commercial and military wooden types in the late 1920's, when the use of radio became widespread, by interconnecting ALL the metal parts with wires or flexible metal tape known as bonding braid. When nails *were* left in the structure they were of brass, which does not build up static charges.

While all sorts of substitute materials were used in the between-wars years, all of the postwar-II designs use aircraft quality materials or substitutes that are acceptable to FAA. Plastics are used in non-stressed areas, while synthetics in the form of fiberglass cloth and Dacron have found wide acceptance as a covering material. They have a great advantage over traditional cotton and linen in that they are not subject to deterioration from the ultraviolet rays of sunlight. Dacron, either from a drygoods store or under the name of CECONITE, developed specifically for aircraft use, has a further economic advantage, because it is shrunk by a hot iron, it saves greatly on the amount of dope needed for the finish.

POWERPLANTS

Except for the very few homebuilts that were standard size and used regular airplane engines from 80 h.p. on up, the post-WW-I designs used a wierd assortment of powerplants. They ranged from converted motorcycle engines to war-surplus 28-h.p. target plane engines, and not a few converted automobile engines. One fine 60-h.p. air-cooled airplane engine, the Lawrence L-4, later Wright "Gale," came out of WW I, but it wasn't on the surplus market and was therefore out of reach of most amateurs.

For those who could afford it, the need for reliable small engines was met in the 1925-1935 period by such European imports as the two-cylinder 32-h.p. Bristol "Cherub" from England, and the lovely little 40-h.p. French Salmson, a baby 9-cylinder air-cooled radial. By 1930, a number of true American lightplane engines had appeared, notably the 26-36-h.p. Aeronca E-107 and E-113 and the 37-40-h.p. Continental A.40 flat-four (which evolved into the famous 65-h.p. A.65.) However, by the time these reliable little mills and their contemporaries became available on the used market, homebuilding in the United States was virtually dead.

Rebirth of the Homebuilts

The rebirth of the homebuilt movement after WW II was notable for the almost 100% use of standard aircraft powerplants, mostly the established 65-h.p. Continental and Lycoming and a few of the later Continental 85's. The 65's were quite easy to come by for a while as many flying schools and rental operators converted their existing 65-h.p. machines to 85-h.p. By the middle 1950's no manufacturer was building a 65-h.p. airplane, so production of this engine virtually ceased and low-time used ones became hard to obtain. This started a demand for an entirely new powerplant source; this was met almost immediately by another European import — the German Volkswagen automobile engine.

The Wonderful VW

At first, this appeared to have everything. It was air cooled, could be converted quite easily to magneto ignition, delivered between 25 and 30 h.p. when fitted with a propeller, and most important of all, spare parts were plentiful and relatively cheap. Americans took to this enthusiastically, and rushed to buy plans for several proven French airplanes that had been developed for it without fully understanding the circumstances under which the designs had been developed. Actually, adaptation of the VW engine to aircraft was a case of European substitution for something that the American had all along. There were no small European airplane engines similar to the American 65's and the cost of importing them for recreational use was prohibitive, so they had turned to an available substitute.

In direct competition with American engines, the VW started with two serious disadvantages—low power and high weight. It weighs almost as much as an A.65, with half the power. This made a very small airplane mandatory, and left little margin for baggage or even gas. And, when first introduced into the country, the VW cost about the same as an A-65. This ratio is changing in the VW's favor as the 65's become scarce, and new American designs are appearing that are better suited to use of this powerplant under American conditions.

The 72-hp McCullough

Two-stroke-cycle engines haven't had any significance in American aviation since before WW I, even in homebuilts, although a very few experimenters have recently tried converted outboards up

The rarest homebuilt design of them all—a two-seat biplane. Access to the front seat is one of the major problems in small biplanes and parasols. Ability of the Duncan "Special" to fly two people on only 65 HP results mainly from generous wingspan. (Photo by Peter M. Bowers, 1962)

Mid-wing monoplane design such as this Stits "Flutterbug" suffers seriously from impaired downward visibility although the situation is alleviated considerably on the ground by use of tricycle landing gear which gives good forward visibility over the nose. (Photo by Peter M. Bowers, 1958)

Although more closely associated with the activities of the Antique Airplane fans rather than the typical homebuilders, full-size replica aircraft are licensed as homebuilts. Major Jim Appleby's 1915 Fokker E-III is accurate even to use of authentic rotary engine. (Photo by Peter M. Bowers, 1962)

to 40 h.p. with some degree of success. In the gyroplane field, however, two-cycle engines enjoy an almost complete monopoly. Most popular is the four-cylinder 72 h.p. McCullough used by the Army in radio-controlled anti-aircraft targets (drones). These are relatively cheap and easy to come by, but have certain inherent disadvantages. They were built to be expendable, and are not up to certificated standards of workmanship. Also their service life is relatively short. These deficiencies are corrected to some extent by firms that specialize in rebuilding the engines for amateur use.

The four-cylinder McCullough, with 100-cubic-inch displacement, weighs only 78 pounds and delivers 72 horsepower. While these figures sound like the answer to a home-builder's dream, there is a catch. The 72 h.p. is delivered at 6,000 rpm through a tiny propeller. This is all right for the 12-foot 200-mph target drones, but will hardly move a standard airplane.

Such high speed and small diameter make an aeronautical propeller highly inefficient so that only the very light gyroplanes and ground effect machines can utilize them. A few powered gliders use the drones, but the propeller inefficiency is largely offset by the greatly increased efficiency of long-span glider wings, which are at least double the length of the average home-built airplane wing. Some experimenters have found that the McCulloughs will fly conventional but fairly long-span ultra-light airplanes by being "de-rated" to 25-30 h.p. through the use of a larger diameter propeller that reduces power by holding the r.p.m. down to 2,500 but compensates for this loss to a large degree through increased propeller efficiency.

Even so, the airplane has to be practically a powered glider. The amateurs found out soon after WW I that the power requirement went up as planes got below about 25-foot wingspan, and that successful flight on 25 h.p. or less was largely a function of light wing loading (pounds of airplane carried by each square foot of wing area) combined with the increased efficiency of long-span wings.

The 125-hp Lycoming

A new 0-290G powerplant has become available in recent years —the 125-h.p. Lycoming ground power unit engine, available as military surplus. This is practically a standard four-cylinder air-cooled Lycoming airplane engine adapted as the power source for running electrical generators and jet airplane starting compressors on the ground. The engines are not immediately usable in aircraft as they have been reduced to single ignition, and a few essential modifications must be made to bring them up to aircraft standard. Since these are not type-certificated engines, the amateur can do this work himself without a mechanic's license.

One disadvantage in buying a surplus engine of this type is uncertainty as to its condition. Some prove to be in excellent shape while others are pure junk, and this cannot be determined until one is taken apart and inspected. As with the drone engines, some firms sell modified and inspected engines to the amateurs. For those who are not in a position to do their own engine work, the additional cost of such a unit is well worth the increased price, which still comes to less than $200, about half the present cost of a good 65.

Because of their weight and power, these ground unit engines are used only in planes near the top of the typical homebuilt range —the heavier two-seaters, aerobatic and fast-climbing biplanes, and the relatively few slick monoplanes that are built primarily for speed.

Auto Engines

Another new and promising source of non-aviation engines is the American automobile industry. Daring breakthroughs have been made with entirely new powerplants starting with the 1960 models, both in standard sizes and compacts. So far, the air-cooled Chevrolet Corvair unit shows the best promise, and several different conversions have already been flown.

Jets: A Distant Dream

The dream of a small jet engine for the homebuilder is still a long way from realization. Entirely aside from the cost, the efficiency drops off so rapidly in small sizes that it is presently doubtful whether a successful small jet can ever be developed. Other handicaps to amateur use are excessive fuel consumption by normal airplane standards and a noise factor that is sure to make them unwelcome at the average small airport.

Since jet efficiency is also a function of forward speed, this means that the planes must be much better aerodynamically than the average homebuilt, flutter-free at high speeds (which is hardly a design problem up to 120 m.p.h.), and in general be a much more sophisticated design than the latter-day antique.

EQUIPMENT

As with the sports cars and small sailboats to which they are most directly comparable, the furnishings and gadgetry associated with amateur-built airplanes are quite austere by industry standards. Certain items are mandatory: engine tachometer, oil temperature and oil pressure gauges, and an altimeter, airspeed indicator, compass, fuel quantity indicator and safety belt. Very few homebuilts use starters. Hand-propping is generally accepted as a natural part of purely sport flying, and the installation of a starter, even when the engine is of a type that can use one, is a severe handicap in a small airplane. In addition to the considerable weight of the unit, there is the need for a generator, battery, voltage regulator, and the rest of the electrical system which can easily add up to a 100-pound weight penalty.

Radio

Radio had been of relatively little use in homebuilts until the FAA-tower requirement of December 1961. Aside from cost, the main deterrents to use of radio had been the need for an airplane electrical system, bonding, and a shielded ignition system. The few radios in use were in the heavier deluxe types or were portable battery sets. The new requirements have already stimulated the electronics industry to develop new low-cost transistorized units, so radio may no longer be considered a special and complicated luxury.

Lighting

Electrical lighting for night flying is a feature very seldom incorporated in homebuilts, which are ordinarily licensed for daytime flying only. With ample justification it may be possible to obtain approval for night operations, but this is hardly a problem for the average owner-builder.

Harness

Shoulder harness is not a required item, as is the safety belt, but it is a very worthwhile investment that costs little and can return whopping dividends in even a minor accident. Shoulder harness is attached to basic structure behind the pilot and comes over each shoulder to fasten to the safety belt and keep him from pitching forward and striking the instrument panel or forward structure with his head during sudden decelerations. The EAA conducts a vigorous campaign for installation of shoulder harness in every member's homebuilt.

Parachutes

Parachutes are seldom used except during test flights and aerobatics. By the time the homebuilt has passed the first few hours of its 50-hour test, it has demonstrated its structural integrity to

Three-quarter-size replicas, such as this Sorrell-Fokker DR-1, combine the romance of the pure antiques with the more practical size, cost, and reliable small modern engines of the typical homebuilt, and are **not considered part of the antique movement.** (Photo by Peter M. Bowers, 1957)

the point where breakup in flight is no longer a consideration calling for the use of a 'chute. In the event of engine failure, most pilots prefer to ride the plane down. The possibility of fire in the air is too remote to be considered, and the size and weight of the 'chute and its accessories are themselves handicaps in the smaller airplanes.

Brakes

Brakes, a rarity on between-wars homebuilts, are, along with steerable tailwheels, in almost universal use today. While not required by Civil Air Regulations, brakes are required by most airport rules—the old-fashioned dig-in tailskids and the later springleaf types are prohibited. The tailskids are not banned because they chew up the turf or scratch the pavement. With the skid unable to dig in and have a strong braking effect on hardtop or pavement, a side gust of wind striking a taxiing plane can weather cock it, and it may hit a parked machine or other obstruction.

This characteristic is emphasized by the fact that, since the general adoption of brakes by commercial aircraft in the early 1930's, the braking function of the original tailskid ceased. The main landing gear was then moved aft a bit for improved directional stability and the weight on the skid, mostly replaced by a steerable wheel, was considerably less. The decreased tail weight contributed to ease of tailwheel steering, but made non-steering swivel tailwheel types even more susceptible to weathercocking. About the only tailskid-and-no-brakes airplanes encountered today are a few homebuilts flown almost exclusively from private dirt or sod airstrips, and the replicas and restored antiques that operate "pure" from established airports. The latter are flown only under special circumstances, and are usually towed to the takeoff point before being started.

Ready-made Components

Whenever possible, homebuilders take advantage of the availability of ready-made components from standard factory-built airplanes, especially in the case of steel-tube engine mounts and the smaller (12-15 gallon) fuel tanks. While the FAA frowns on the use of cut-down "major" components such as wings, tails, and fuselages, considering this practice contradictory to the intent of the amateur-built category, it does permit the use of mounts, tanks, cowlings, and sometimes even complete landing gear assemblies. These are specialty items that are highly important to safety, and in some cases even the airframe manufacturers have them made elsewhere by specialist firms.

In the between-wars years, propellers for homebuilts were practically all of wood, many of them homebuilt to go with non-standard engines. Since WW II the props are mostly factory-made metal types. Wood props in the 65-85 h.p. range are hard to find since they are no longer standard production items and can't be picked up at surplus prices as they were right after the war. A wooden prop of just the right pitch/diameter may have to be made specially by the few shops that still do this work.

Hobart Sorrell, of Rochester, Washington, demonstrates the light weight of his all-wood biplane, which features a de-rated 40 HP Mercury outboard engine, racing sulky wheels, and "geodetic" construction. Structure shown here weighs 75 pounds and overall wingspan is 20 feet. (Photo by Peter M. Bowers, 1959)

The metal designs have many advantages, notably the ability to be twisted to the correct pitch at any propeller repair shop and to be straightened out if bent in a minor accident that would shatter a wooden prop. Another big advantage is that it is possible to use metal props of slightly lesser diameters than are required for production planes. Repaired props that end up just too small for the standards become available for the homebuilts at bargain prices. Further, metal props never get out of balance and are not subject to deterioration when the plane is tied out in the weather. Controllable-pitch props, which have been on the market since just before WW II in lightplane sizes, simply aren't justified on most homebuilts. Even the lightplane industry sticks with fixed-pitch under 175 h.p.

All other things like extra instrumentation, ashtrays, and fancy upholstery are purely items of pilot preference. Some derive great satisfaction from dolling up a perfectly standard design to the point where it is highly individualistic, while others put in no more than the bare minimum needed to get it into the air legally.

SPECIAL FEATURES

Like other extras, this is also a field of pilot preference, with the advantages and disadvantages to be weighed against the "regular" configurations. One of the more popular features is a single-leg spring steel landing gear in place of the more conventional tripod units incorporating oleo-pneumatic shock absorbers

The two-seat Lacey M-10 illustrates a unique approach to the folding wing problem. See pages 111 and 113 for other applications. (Photo by Peter M. Bowers, 1962)

or rubber cord "Bungees." This unit was developed by Steve Wittman of Oshkosh, Wisconsin, for his backyard racers of the 1930's, and was introduced into the commercial market on the Cessna airplanes of 1946.

These contribute greatly to the appearance of a clean design, but in spite of their simplicity they usually carry a stiff weight penalty and require an extra-strong supporting structure. These can be custom-made by various shops, or homebuilders can make their own from spring steel stock and then have it heat-treated.

They can also cut down the heavier Cessna units. Some save weight by using tempered aluminum. In the interest of lightness and simplicity, some homebuilts do not use shock absorbers at all, relying entirely on low-pressure tires, usually 8.00 x 4's, for the job.

Folding wings to simplify the storage problem have never been considered very seriously for commercial planes, or even for homebuilts, until the EAA design contests of 1960 and 1962 (see Chapters 10 and 11). The necessary fittings and mechanism add weight and structural complexity, and then there is the problem of disconnecting a number of different things and actually folding the wings. An even more important consideration is trying to do it without help, and then making sure that all the connections are safetied before flight. One minor disadvantage of most folding wing designs is that the entire weight of the wings moves aft, greatly increasing the weight on the conventional tailwheel and making it difficult for one man to lift the tail to move the machine, or raise it to connect a trailer hitch.

Wing flaps add considerably to the structural and aerodynamic complexity of a design, and are used mainly as a crutch to produce

Sorrell-Robinson "Cool Crow" is an experiment in ultra-light construction combined with light span loading for flight with 30 HP two-cycle target drone engine. (Photo by Peter M. Bowers, 1961)

additional lift during takeoff and landing. Their purpose is to compensate for such aerodynamic shortcomings as a simple deficiency in wing area or the excessive wing loading of a small two-seater.

Enclosed Engines

Enclosed engines are largely a matter of builder preference if they are optional on a purchased-plans design. The majority of cowlings use a stamped metal front plate from a standard airplane, with flat-wrapped hinged sheet aluminum panels behind it. Fiberglass cowlings can be made to more complex shapes for special installations, but are heavy because of the stiffness requirement, expensive in material, and offer a terrific amount of work in the making of suitable moulds. Fiberglass or hammered metal wheel fairings called "pants" are used mainly for looks on the slower types, since they have little effect on performance at speeds under 120 m.p.h. Other than the "Schmaltz Factor," the work of building them is hard to justify for anything but a really fast ship. Some firms have standard sizes available at reasonable cost.

Some builders who do adopt the appearance and streamlining advantages of the enclosed engine go a step farther in search of increased performance by using thrust-augmenter exhausts, which add the pressure of engine exhause gasses to the propeller thrust. Cooling is also improved by the increase of cooling air drawn inside the cowl by the venturi action of the concentric arrangement of the exhaust stack discharging high-velocity gas through the augmenter tube, which has its inner end open to the inside of the cowling.

4. Sources of Designs and Material

There are plenty of proven designs available to the home builder who prefers to buy someone else's plans rather than work up his own. The most popular models, with prices and sources of the blueprints, are listed at the end of this chapter and are illustrated throughout the book. Ready-built airplanes on a production basis are NOT available. The privileges of amateur-built airplanes in the Experimental category apply only to true "homebuilts." See Chapter 8 for other legal considerations.

The major problem for the customer in picking a design lies in making the right selection, especially if he is not already a pilot and doesn't really know what type of airplane is best suited to his actual or anticipated needs. Too often such a person chooses his design purely on the basis of looks or an idealistic concept of what he thinks he'd like, with eventually disappointing results. The inexperienced, then, should talk over their selctions with experienced friends, being careful not to be swayed by that friend's own enthusiasm for some machine that is particularly suited to himself and not to the person consulting him.

Most of the designs for which do-it-yourself plans are available are advertised in *Sport Aviation,* national publication of the EAA, and to a lesser degree in the general aviation and mechanical magazines. The law of "survival of the fittest" doesn't always apply in this field since the machines are not directly competitive and there is a wide range of personal preferences to meet. Unfortunately, some marginal designs manage to get by on almost pure eye appeal and schmaltz, while others which are short on looks are really fine flying machines. The drawings range from barely adequate to a surplus of meticulous detail, but in practically all cases designers assume a certain amount of knowledge on the part of the builder in the use of tools and aircraft fabrication procedures.

Since kits containing significant quantities of pre-fabricated parts violate the intent of the amateur-built category by taking some of the work away from the builder, full kits are not available for machines intended to be certificated as amateur-built. Complete kits are on the market for several machines, both gliders and airplanes, but these are fully type-certified designs and are licensed in the standard category, even when built by amateurs.

The only complete kits on the market are for aircraft that are licensed in the Standard category rather than as amateur-built, such as this Schweizer 1-26 sailplane. Much of the real work is done on the kit—the steel tube fuselage is welded, the wing spars are riveted together, and all metal parts are stamped out. (Photo by Peter M. Bowers, 1957)

For those who want to do their own designing, the EAA has issued several publications that cover various aspects of design, materials, and construction. Several individuals who have worked up "Design Manuals" directed especially toward amateur design also advertise in *Sport Aviation*. The Government Printing Office, too, publishes many documents on aircraft construction, maintenance, and operation. The most important of these for the home builder, although somewhat misleading in title, is FAA Manual 18, "Maintenance, Repair, and Alteration of Airframes, Powerplants, Propellers, and Appliances." Its value lies in its presentation of all the approved fabrication methods for all conventional types of aircraft structure, and no serious home-builder should be without it. Write to the Superintendent of Documents, Washington 25, D. C., for a listing of other government aircraft publications.

Most public libraries will have aeronautical engineering textbooks that will be of some help. However, completely designing an entirely new airplane from scratch with the aid of textbooks is an almost impossible task for an inexperienced amateur. These books are not oriented to the airspeeds, weights, and load factors in the typical home-built range, and the would-be designer does not have a backlog of experience on which to base the numbers that he needs or to correct the "big airplane" figures.

Actually, most amateur designing (and some professional) is by the "copycat" procedure. General layouts and proportions follow existing designs of proven performance and safety, and structural

One plentiful source of present-day designs is the past. Many inadequate but still appealing designs are relatively easy to redesign with modern features and improvements. Compare this greatly-improved Flying Flea with the original 1935 version on Page 13. (Photo by Peter M. Bowers, 1960)

component sizes are largely chosen on the basis of: "Such-and-such an airplane is in the same weight and speed class as mine. It has spars one inch thick, so I'll use the same and maybe add an eighth." With this technique, a very successful and conservative design can be developed merely by averaging out the proportions of a number of proven designs of the type desired.

Finding suitable materials from which to build an airplane used to be a problem. Now, thanks to the popularity of the movement, there are many specialized firms that cater to the special needs of the homebuilder. Many advertise in *Sport Aviation, Trade-A-Plane News,* and the general aviation magazines, and a few that cover a wide range of materials are listed at the end of this chapter.

Salvaging Junked Parts

Home-builders who live in or near those cities in which commercial and military aircraft are manufactured are very fortunate, for large quantities of raw material such as sheet aluminum, fittings, steel tubing, nuts and bolts, and countless other goodies are constantly showing up in the local salvage and junk yards where they can be bought at a fraction of their new cost. People in other areas must buy these items from firms that advertise in the trade publications or from parts suppliers at the local airport.

The junk pile of the local airport can also be a good source of miscellaneous material. Sometimes complete wrecked airplanes, less engine, instruments, wheels, and other components of value to the operator, can be bought for a few dollars. These are good sources of tubing, bolts, pulleys, and sheet metal that can be cut down to smaller units. One disadvantage of this source is that it may not provide quite enough of a certain material for the job, with the result that a lot of time is wasted in finding the needed amount to match. Useful used items like wheels, engines, instruments, fuel tanks, and propellers can also be found at the airports, but not on the junkpile. These are the first things salvaged from any wreck.

Engines, unfortunately for the home-builder, are very sensitive to the laws of supply and demand. The recent popularity of homebuilts has run the demand for flat-fours from 65 to 85 h.p. way up, while the supply is steadily diminishing. Sixty-fives went out of regular production several years ago when the aircraft industry went to higher power for the lightest production airplanes.

Finding a Prop

Finding a suitable wooden propeller for a homebuilt is quite a problem. A single-seater with a 65-h.p. engine is apt to be quite a bit faster than the Piper Cub or Taylorcraft that the engine came out of, so the prop that was on the engine, matched to the standard r.p.m. of the engine and the speed of the original airplane, will be pitched too low for the homebuilt. A prop from an 85-h.p. engine in another production airplane having the same speed as the homebuilt won't do either because of entirely different engine

Some oldies need relatively little modification to look thoroughly modern. Jack McRae, of Hempstead, L. I., redesigned the Driggs "Dart" lightplane of 1925/26 into the "Super Dart," the principal outward changes being spring steel landing gear and a fully-enclosed Lycoming engine. (Photo by Howard Levy, 1955)

speed. When one must use a wooden prop, it will very likely have to be made by a shop that turns out custom-made props. Wooden ones have not been in regular production since shortly after World War II. The majority of home builders use production forged metal propellers that any approved propeller repair shop can twist to the exact pitch required for the particular airplane.

Dope and fabric can sometimes be bought in relatively small quantities from operators at the local airport, but if you need significant quantities you will usually have to order from regular supply stores or advertised sources. Dacron fabric suitable for use on non-type-certificated aircraft can be obtained from most local dry-goods stores under the name of Dacron Polyester Taffeta.

Aircraft-grade plywood has practically vanished from the American scene, and most of that used by the homebuilders is imported from abroad. See end of chapter for sources. Marine plywood, acceptable for use in home-builts and available in thicknesses from $1/8''$ up, is relatively easy to come by in boating areas, but often has to be ordered through the local lumberyard in areas where it is not normally stocked. Boat or ladder spruce is generally available through lumber yards, and sometimes stacks of this material yield good spars, but as a general rule it is best to order aircraft grade spruce as such for spars from the firms that specialize in it.

The local hardware store will provide some materials, but not very many; spar varnish and weldwood glue, mainly. Regular hardware store nails and stove bolts will not do. The nails must be cement-coated, parkerized, or of brass, and the bolts should be AN (Army-Navy) standard with elastic stop nuts or castellated nuts and cotter pins instead of coarse thread bolts with square nuts and lock washers. These are available at the local airport in some cases, or from established aircraft supply houses along with the steel cable, control pulleys, and fittings that should be only aircraft grade. If local aircraft stores do not carry S.A.E. 4130 steel tubing and sheet, it can be obtained on order through local metal supply houses.

Turnbuckles and shackles in sizes suitable for the control systems and wire bracing of homebuilts are becoming increasingly difficult to find. Only a careful check of the advertisements or some real digging can turn up these items. Some surplus yards still have a few old-fashioned streamline wires, but chances are

An entirely original design developed to meet specific requirements of the builder alone. Hobart Sorrell's ultra-light Model DFG, developed for experimental flying with a converted outboard motor. (Photo by Peter M. Bowers, 1959)

they won't be the right size. For biplane rigging and the internal drag bracing of two-spar wings, 1 x 19 stainless steel stranded wire is ideal. It is obtainable through various hardware and machinery suppliers, and can be installed either with swaged sleeve fittings or with Nicopress fittings. Some builders eliminate the turnbuckle problem for drag bracing by using $\frac{1}{8}$" to $\frac{1}{4}$" steel rod threaded at each end; they either install fork fittings on the ends or poke them clear through the spars, anchoring them on angle blocks.

Aircraft Plans

The plans listed below are only a few of the many that are available to the homebuilder. This listing has been limited to those designs which have been on the market for some time and have proven their dependability through wide use or have otherwise demonstrated their capabilities to qualified judges.

Designation	Source	Cost	Remarks
Baby Ace "D"	Ace Aircraft Mfg. Co. c/o Ed Jacobs, McFarland, Wisconsin	$20.00	1-seat parasol mono, wood wing, steel tube fuselage, 65-86 h.p. (see Chapt. 2).
Bensen "Gyro-copter"	Bensen Aircraft Corp. Raleigh-Durham Airport, Raleigh, N. C.	data $2.00	1-seat autogyro with 2-cycle drone engine (see Chapt. 3).
Bensen "Gyro-glider"	As above	data $1.00	1-seat rotary-wing kite or free-flight glider EAA design contest winner.
Bowers "Fly Baby IA"	Peter M. Bowers 13826 Des Moines Way Seattle 88, Wash.	$15.00 Brochure and photo, $1.00	1-seat all-wood low wing monoplane (see Chapts. 10 & 11). 1B uses biplane wings.
Brigleb BG-12A	Wm. G. Brigleb, El Mirage Field, Adelanto, Calif.	Kits only	1-seat all-wood high performance sailplane.
Commuter Jr.	Helicopter Research Co. Box 121, La Miranda, California	$35.00	1-seat steel tube helicopter, 65-90 h.p.
Druine "Turbulent"	Falconar Aircraft Municipal Airport Edmondton, Alberta, Canada	$40.00	1-seat all-wood monoplane for VW engine
E.A.A. Biplane	Experimental Aircraft Assoc. 9711 W. Forest Park Drive Hales Corners, Wisc.	$10.00	1-seat biplane, wood wings, steel tube fuselage, 65-90 h.p. (see Chapt. 11).
Hannaford (Rose) Parakeet	Foster Hannaford Jr. Mundelein, Illinois	? ? ?	Improved Rose 1-seat biplane for 65-85 h.p. (see Chapt. 1).
Honey Bee	Bee Aviation Inc., 1536 Missouri St. San Diego 9, Calif.	$35.00	1-seat all sheet metal monoplane, type certificated. 65 h.p.
Jodel D-9	Falconar Aircraft Municipal Airport Edmondton, Alberta, Canada	$30.00	1-seat all-wood low wing monoplane, VW or 65 h.p. Cont. engine.
Meyer "Little Toot"	Meyer Aircraft 211 Seminole Trail Pensacola, Florida	$50.00	1-seat biplane, sheet or tube fuselage, 90 to 150 h.p. (see Chapt. 11).
Midget Mustang	Mustang Aircraft Co.	data $2.00	1-seat low wing sheet metal monoplane 85 h.p.
Mong "Sport"	Ralph E. Mong, Jr., 1218 No. 91st East Ave., Tulsa 15, Oklahoma	$39.00	1-seat biplane, steel tube fuselage, 65-85 h.p. (see Chapt. 11).
Nesmith "Cougar"	Robert E. Nesmith 6738 Long Drive Houston 17. Texas	Believed to be $6.00. Write for confirmation	2-seat steel tube fuselage, wood wing, 85-125 h.p.

Designation	Source	Cost	Remarks
Nesmith-Eaves "Cougar"	Leonard Eaves Oklahoma City, Okla.	? ? ?	3rd place, EAA Design Contest. Folding wings on standard Cougar. (see Chapt. 11).
Schweizer 1-26A	Schweizer Aircraft Corp., Box 147, Elmira, New York	Kit or finished machine only. Write for current prices.	1-seat all-metal ATC'd sailplane.
Schweizer 2-22	As above	As above	2-seat all-metal training sailplane.
Skyhopper	Skyhopper Airplanes Inc., 2445 South Beverly Drive Los Angeles 34, Calif.	$50.00 data $2.00	1 or 2-seat mono, wood wing, steel tube fuselage. 65-85 h.p.
Smith "Miniplane"	Mrs. Frank Smith, 1938 N. Jacaranda Place Fullerton, Calif.	$25.00	1-seat biplane, wood wings, steel tube fuselage, 65-125 h.p. (see Chapt. 11).
Smith "Termite"	Wilbur Smith 1209 N. Rosney Bloomington, Ill.	$15.00	1-seat all wood parasol monoplane, 36-65 h.p.
Sportaire II	Al Trefethen 2432 Chapman St., Lomita, Calif.	$70.00	2-seat side-by-side low wing monoplane, wood wing, steel fuselage, tricycle gear 65-90 h.p.
Stits "Flutterbug"	Ray Stits P. O. Box 2084, Riverside, California	$35.00	2-seat tandem mid-wing monoplane, wood wing, steel fuselage, tricycle gear 65-90 h.p. (see Chapt. 3).
Stits "Playboy"	As above	$25.00	1 or 2-seat low wing monoplane, wood wing, steel fuselage, 65-150 h.p. (see Chapts. 2 & 11).
Stits "Skycoupe"	As above	$45.00	2-seat side-by-side cabin monoplane ACT'd. wood wing, steel fuselage, tricycle gear.
Stolp-Adams "Starduster"	Stolp Aircraft P. O. Box 461, Municipal Airport, Corona, California	$25.00	1-seat biplane, wood wings, steel tube fuselage, 85-125 h.p.
Turner T-40	E. L. Turner Fort Worth, Texas	? ? ?	2nd Place, EAA Design Contest. All-wood low wing monoplane, 85 h.p (see Chapt. 11).
Volmer "Sportsman"	Volmer Aircraft 104 E. Providencia Ave. Burbank, California	$125.00 data $2.00	2-place side-by-side flying boat designed around Aeronca Chief or Champ wings & tail.
Wittman "Tailwind"	S. J. Wittman P. O. Box 276, Oshkosh, Wisconsin	$125.00	2-seat cabin monoplane, wood wings, steel tube fuselage, 85-125 h.p. (see Chapt. 11).

Note: See List of Aircraft Material Sources at end of book

5. Original Design Procedure

Designing an airplane is a really complex job if one actually creates an entirely original but still conservative design, and goes through all the procedures of aerodynamic analysis, airfoil selection, stress analysis, material size determination, and so on. Merely describing these various steps would fill a volume several times the size of this guide, and those who are capable of doing the job have training that puts them far beyond the point of needing this document.

For those who are content merely to produce a distinctively personal flying machine within the well-proven envelope of "average" airplanes, the following general rules will enable even the unskilled to develop an acceptable airplane by the "Copy-Cat" technique. Realize, however, that such a ship will be no world-beater. Its prime function will be to provide the designer-builder with a safe recreational machine that is distinctive because it is "his own" and not a stock model. The pilot's principal satisfaction comes mainly from being off the ground in his own creation rather than in having a ship that is a few miles per hour faster than someone else's.

This is sometimes a hard point to get across to pilots who have flown only production machines, especially those who are not yet pilots but who want to build a homebuilt. Both have the disadvantage of not having a "feel" for sport airplanes or their capabilities and limitations. Generally, they expect too much — such things as 130-150-m.p.h. cruise and a 50-m.p.h. landing speed, four hours' range, and acrobatic performance, all with an 85 h.p. engine. They can usually end up with any two, or maybe three, of these characteristics, but never all of them at once. As they become accustomed to the fact that sport planes are an end in themselves, rather than being competitive equipment, they begin to see that some of these desirable features, originally so important, are not really significant.

Of course, practically any conservative design can be altered to achieve certain desired performance goals, but the designer-builder should take a long and careful look at what this particular performance is costing him. The cost usually shows up as a loss of other desirable characteristics. One way to find out whether or not a change is justified is to use the aviation industry procedure known as a "Trade Study." This consists of making up a list of two columns, one listing all the gains expected from a particular feature or change from standard, and the other listing the disadvantages and the costs in terms of effect on other characteristics. If the trade ends up with more bad features than good, a re-evaluation is in order.

Want More Speed?

A good example is a desire for more speed. There are several ways of getting this, the easiest being to install a more powerful engine, say a 125-h.p. model instead of an 85. The first and most obvious disadvantage is increased weight. This automatically results in increased landing speed, which can be countered by adding flaps, but there are other less obvious disadvantages. Since the engine is in the nose in most designs, an increase in weight will upset the balance and call for a rearrangement of internal equipment—or for ballast in the tail, and more weight.

A heavier engine will usually require a larger, stronger, and *heavier* engine mount, and structural changes to the fuselage may be necessary to accommodate it. A new cowling will probably be needed, and certainly a new propeller. Finally, in order to maintain an adequate cruising range, the fuel capacity will have to be increased, calling for a new tank, different tank installation fittings, larger fuel lines, still more weight, and so on. It is up to the builder to decide whether an extra 20 m.p.h. is worth all this.

The above example was chosen because of significant differences in weight and size between the commonly-used 85-h.p. Continental engine (188 cubic inch) and the 125-h.p. Lycoming (280 cu. in.) Changes from the 65-h.p. Continental (170 cu. in.) to the 85-h.p. are not nearly as involved, as there is no size change and very few pounds in weight. Changes in this lower power range are seldom made for speed, but rather for improved takeoff and climb characteristics, the principal deterrent being the cost or availability of the more powerful (and sought-after) engine.

A less costly way of increasing speed is to clean up the design, using a closed cowling around the engine, putting "pants" on the wheels, and adding streamline fairings to strut or wire terminals. Still another method, famous in air racing history but actually an aerodynamic fallacy, is to clip the wings. Theoretically, this increases speed by cutting down drag, which it does to some extent.

What it really does, in order to lift the same load with less area, is to cause the airplane to fly at a higher angle of attack in order to get an increase in lift coefficient to match the area loss. The drag INCREASE from the higher angle more than offsets all the anticipated gains of clipping, while the normal landing characteristics go to pot because of the increased wing loading and the decrease in wing efficiency due to shortening the span and the effective aspect ratio, both of which are very important to the low-speed characteristics of the plane.

Another speed-increasing procedure, which must be taken before construction gets underway, is to redesign the wing to use a "faster" airfoil section. However, this really puts the builder on thin ice in several directions by introducing all sorts of changes in load distribution, setting of the wing relative to the fuselage, and in some cases making an entirely different airplane of it.

The Importance of Low Speed

For normal sporting purposes, good low speed characteristics are far more desirable than an extra 10 or 15 mph at the other end of the scale. Because of their small sizes, which result in low Reynolds Numbers that give a small single-seat monoplane grossing 700 or 800 pounds an effective wing loading equal to that of an early WW II fighter plane, even the lightest homebuilts land quite "hot" compared to a Piper Cub, Aeronca Champion, or other standard production "floater." Anything that will INCREASE the landing speed is to be avoided whenever possible. Many builders have even undertaken "anti-speed" programs to improve the landing, takeoff, and climb (generally referred to as "Over-the-Fence Characteristics") of available designs by adding a couple of feet of wingspan at the expense of the high speed, which they had finally recognized as being of little actual importance. A good historical example is the evolution of the Heath parasol, a famous homebuilt design of 1927-1932. This originally had a wingspan of 25 feet, but it grew to 30 before the plane could demonstrate flight characteristics that enabled it to win an FAA Approved Type Certificate.

In general, those who are building an airplane for the first time from the plans of an established design will do themselves a great favor by avoiding the temptation to make improvements. If the original designer had considered them worthwhile, he would have added them himself. Similarly, in developing an original design for the first time, one should not depart too far from the average proportions, weight/power distributions, etc., of equivalent designs unless experience shows that the deviation is practical, safe, and not too costly.

In many cases, "freak" airplanes have been built simply to test an invention that someone dreamed up. There are countless examples back through aviation history to show that an inventor, who has come up with what might have been a sound idea, was unable to prove it out because the airplane he built departed so far from the norm, either by his own intention or merely because he didn't know enough to design it properly in the first place, that it wouldn't work well enough to give the invention a chance to prove itself. If a homebuilt is to be used as a test bed it should by all means be as conservative as possible so that the effectiveness of the device under test can be measured against a relatively normal performance.

The Big Choice

Designing an airplane involves the biggest collection of compromises you could possibly imagine. Practically every step in the whole procedure requires a decision between what is desirable and what is practical, and the process begins before the first line ever goes down on paper. Just deciding what type of plane to design is a problem, and too often a person unfamiliar with sport planes picks a design that he *thinks* will suit him when it actually doesn't at all. In some cases, the decision is made for political, rather than technical, reasons. The ship might have to be a two-seater on the grounds that hubby isn't going anywhere if wifey can't go too.

Monoplane, biplane, replica, racer, flying boat or roadable airplane; wood, steel, tubing, or sheet metal; open cockpit or cabin; pusher or tractor — all must be evaluated and call for decisions! Decisions! Decisions!

DESIGN PROCEDURE

Once the configuration and structure have been chosen, make a survey of equivalent designs to get a feel for the actual proportions, locations of parts, and sizes of the structural members. If possible, make up a table listing several models and such data

Weight and Balance

Since the pilot and the powerplant between them make up nearly half the gross weight of a small airplane, and their weights are not subject to compromise, the design of the plane will have to be adapted to them. If the average proportions of successful airplanes in the same size/weight class are closely adhered to, the designer should have little trouble. However, even when following standard proportions, a mathematical check is in order.

Use the method shown in FAA Manual 18, and pinpoint the weight of every item of fixed and movable equipment. The loaded Center of Gravity (CG) should come out somewhere between 20 and 30 percent of the Mean Aerodynamic Chord (MAC) of the wing. There's no room for compromise here; for the airfoils in use today, the CG *has* to be within those safe limits or the plane will have dangerous characteristics. Juggle the location of the components, including the pilot, or add ballast to get the CG location.

The engine is a fixed-weight item and stays put. The pilot's weight is a variable item and also makes up anywhere from 20 to 25 percent of the gross weight of a small single-seater. To avoid drastic changes in trim when changing from a 150-pound pilot to a 200-pounder, the seat should be right on the CG so that a 50-pound weight change won't affect the longitudinal balance. The crew/weight problem is especially acute in tandem-cockpit two-seaters. In the old production designs of this type, the passenger's cockpit was right on the CG, but in the smaller home builts the CG usually ends up between the two, with big balance changes resulting from a full or empty second seat. The easy way out is to design a side-by-side two-seater, but the fat fuselage needed to seat two large people introduces *many* structural and aerodynamic problems while solving only one.

Fuel is also a variable load, starting with anywhere from 12 to 25 gallons at six pounds per gallon at takeoff and burning most of it off before landing. Because of this weight change, which averages about 12 percent for most planes, the fuel tank should also be right on the CG in order not to affect the trim. This is possible only in planes using wing tanks.

A modification of an already established design, the Wittman W-9L, developed from the well-known W-8 version by Steve Wittman himself, the major changes being a 150 HP Lycoming engine and a unique tricycle landing gear installation. (Photo by Peter M. Bowers, 1960)

Wing tanks with gravity feed can be used on high-wing monoplanes with wings thick enough to accommodate them, but complexity is introduced on low-wing monoplanes by the need for a fuel pump to get the gas up to the engine. When the tank is installed in the fuselage, a satisfactory compromise must be achieved between tank and pilot locations. Without actually knowing the weights of the structural items, a rough idea of the gross weight of a typical small homebuilt design can be determined from the known weights in the following formula:

$$\text{Gross weight} = \frac{\text{useful load} + \text{complete engine installation}}{.60 \ (.56 \text{ to } .64)}$$

Useful load consists of pilot, fuel, and baggage, and offers some variables that can be controlled, while engine weight is a fixed value. Baggage is usually the variable that suffers most, and while some homebuilts have space for baggage, relatively few can afford the weight penalty.

The diagram on page 59, worked up by George McVey of EAA Chapter 26 and an aeronautical engineer at Boeing, presents average proportions that have proven satisfactory for homebuilts of conventional configuration. While a high-wing or parasol monoplane is shown, the diagram applies equally well to low-wing monoplanes and to biplanes. For the latter, the mean aerodynamic Chord of *both* wings is established and then treated as though the ship were a monoplane. The numbered items are elaborated upon in the following text.

DETAIL CONSIDERATIONS

Wing Design

This is the most important single unit on the airplane, and is subject to the widest choice of design variation. Some gliders and airplanes are all wing and nothing else. The style and location of the wing will have been determined by the initial choice of the airplane that is to be built, which for today's homebuilts means that roughly four out of five choices will be for a monoplane.

The vertical location of the wing has little effect on actual performance in sport-type airplanes, and is primarily a matter of pilot preference. The parasol location has the advantage of excellent pilot visibility downward, but undoes it all by blocking the upward view completely when the pilot is seated right on the CG. It is also a bit of a handicap structurally because of the need for a complex strutting system to hold it above the fuselage, and the access problem can be quite acute when a big pilot tries to squeeze into an open cockpit directly under the wing.

The high wing, with the spar fittings attached to the upper longeron, is one of the simplest structural arrangements, especially for a strut-braced wing on a cabin-type fuselage. Midwings, with the wing located somewhere between the upper and lower longerons, are relatively rare in sport planes but are quite common on the midget racers. Primary disadvantages are squeezing the pilot between the wing spars and extremely poor visibility downward from the cockpit.

The low-wing configuration is the most popular for homebuilts, primarily because of the excellent all-around visibility offered and easy access to the open cockpit by being able to use the wing as a step. On some designs, the wing supports the landing gear. Strut-braced two-piece low wings are at a bit of a disadvantage because of the need to locate the struts above the wing as compression members. The strut/wing fittings and the acute angle between strut and upper wing surface destroy quite a bit of useful lift in the immediate area.

The Biplane

In commercial aviation, the biplane has been a dead duck since the middle 1930's. It offers a distinct advantage over the monoplane only in maneuverability because of being able to

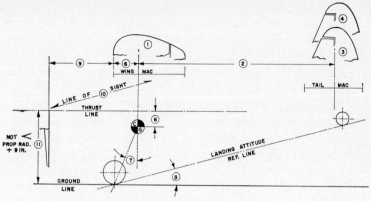

(1) Total Wing Area = $\dfrac{\text{Gross Wt}}{\text{Wing Loading}}$ (Not > 18)   AR = $\dfrac{\text{Span}^2}{\text{Area}}$,
  SPAN = $\sqrt{\text{AR}(\text{AREA})}$ (AR = 6 to 7)

(2) Tail Hinge Line LOC = 2.5 to 3.5 $\left(\dfrac{\text{Wing}}{\text{MAC}}\right)$, AFT of CG

(3) Total Vert. Tail Area = $\dfrac{.034\,(\text{Wing Span})\,(\text{Wing Area})}{3\,(\text{Wing MAC})}$,
  AR Not < 1.5

(4) Total Horiz. Tail Area—Where Semi-Span = 16.5% of Wing Span, & AR = 4 ∴ AREA = $\dfrac{\text{Tail Span}^2}{4}$

(5) Horiz CG Loc = 25 to 30% Wing MAC Aft of LE of Wing & near CL of A/P

(6) Vert CG Loc should be No > 20% of Wing MAC Above or Below Thrust Line

(7) Main L.G. Loc should be Not > 20° Ahead of & Not < 12° Ahead of CG

(8) Landing Angle = 90% of CL Max (Not < 15°)

(9) Not < Prop Radius

(10) Parallel to (8)

incorporate a given wing area in a much shorter wingspan. In speed for a given power, range, maintainability, and all of the other items that make dollar sense to commercial operators, the biplane is inferior to the single-winger. Only in the specialized field of dusting/spraying does the biplane hold a commercial advantage. These considerations do not apply in recreational areas, however, where the two-wingers have wide appeal. All of the top acrobatic ships are biplanes and two wings, struts, and wires provide a nostalgic link with "the good old days" of aviation that the monoplane cannot match. Because of this fact alone biplanes will survive as long as there is recreational flying.

The First Decision

The first decision to make in wing design is the area necessary to carry the full load of the plane. Structurally, a wing of low aspect ratio (AR), the ratio to total span to Chord (Chord is the distance from leading to trailing edge), is the most desirable. Aerodynamically, however, the higher the AR the better. The higher the AR the less the induced drag, and since induced drag is inversely proportional to speed, low AR is a serious handicap to the desirable low landing speeds. AR's of 6 to 7 are typical for sport airplanes, while they start at 10 for utility gliders and go as high as 24 for the highest performance sailplanes.

On some midget racers they are as low as 3, but these are special-purpose machines in which such luxuries as good landing characteristics are willingly sacrificed.

What Type of Construction?

The second major decision to make in wing design is the type of construction; whether single unit or two-piece, cantilever or externally braced. Two-piece units are by far the favorites, and are much easier to handle in the shop and to install on the airplane. Cantilever wings present special problems of their own. With no external bracing, the wing-to-fuselage attach fittings are very critical, taking all of the flight loads to a safety factor of 8 on some types with no looseness or play at all. Further, the wing structure itself must resist the torsion loads that tend to twist it.

The easiest way to handle these is to cover the wing with a rigid skin of sheet metal or plywood. Most gliders and a few powerplanes handle it by using a single spar at about the 33% chord point and wrapping plywood or metal around the leading edge from the top of the spar to the bottom, forming a torsion-resistant "D-tube" structure. The area aft of this single spar can be fabric covered. Some two-spar cantilever wings are fabric covered but use elaborate internal trussing to take the torsion loads. Torsion is taken by the struts or wires on externally-braced wings, which can be built much lighter because of the external load-carrying features and the ability to use simple pin joints at the fuselage fittings.

On cantilever wings, the drag loads must also be taken out by the skin on a D-tube spar, and transmitted to the fuselage through an additional drag fitting. On two-spar externally-braced wings, the drag load goes into the fuselage through the rear spar fitting after being distributed throughout the wing by the single-plane internal drag wires or tie rods.

Straight or Tapered?

There are many sides to the question of using straight or tapered wings. The straight ones have the advantage of being somewhat easier to build because most of the ribs are the same size and the rib/spar intersections are all right angles. Further, there are slight aerodynamic advantages in that straight wings tend to stall at the root first, while tapered wings stall at the tip first, to the detriment of aileron effectiveness at this critical time.

The tip-stall problem is easily licked by "washing out" the tip, that is, decreasing the angle of attack. On some high-performance sailplanes with sharply-tapered wings, this wash-out, properly called "aerodynamic twist," is as high as six degrees. To assure good aileron control through the stall, even some straight-wing types like Piper "Cubs" wash out the wingtips.

Leading edge slots at the wingtips are usually crutches added as a last resort to lick a stall problem resulting from some design deficiency. The shorter tip chord of the tapered wing reduces the Reynolds Number for a given airspeed, slightly reducing the efficiency of the wing. Advantages of taper are spar cross-sections and wing area distribution with a more direct proportional relationship to the actual load distribution across the full span of the wing, and a slight reduction in rolling moments due to concentration of the mass of the wing closer to the centerline of the airplane.

Taper ratio, the ratio of the root chord to the tip chord, seldom exceeds 2:1 for a homebuilt, and is more apt to be a lesser ratio like 3:2. Determining the MAC of a tapered wing can be done graphically as shown in the following sketch:

Add the root and tip chords as shown. A diagonal connecting the extremes and intersecting the median chord will establish the mid-point of the MAC.

Tapered wings can be had in any combination; straight leading edge and swept-forward trailing edge, or vice versa. The favorite seems to be symmetrical taper about the 25% chord point. The 25% chord line runs straight from wingtip to wingtip and the leading and trailing edges sweep toward it from the root outward. Most tapered wings involve two different tapers, the most

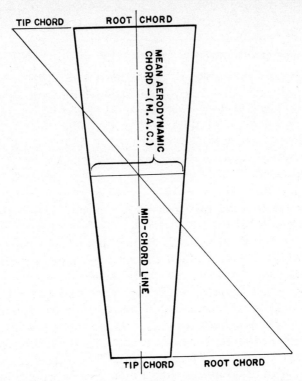

common being taper in planform only, where the tip is of less chord than the root.

If the airfoil used has the same thickness relative to the chord at both the root and the tip, there is no taper in thickness even though there is an actual dimensional difference between the two thicknesses. If the root airfoil thickness is 15% of the root chord, while the tip airfoil is only 10% the thickness of the tip chord, the wing (aerodynamically) has taper in thickness. Most taper-winged homebuilts do without this, although it is a good way to avoid a need for wash-out by using a tip section that stalls at a higher angle of attack than does the root.

Sharp sweep-back of the whole wing has its purpose on super-sonic airplanes, but is useless for any aerodynamic reason on a homebuilt. It is justified on some otherwise straight-winged monoplanes to a small degree to relocate the MAC to conform to a poorly-located fuselage CG. Major distavantages are the structural ones of putting the wing loads into the fuselage through angled fittings and a need for rib-to-spar joints not at right angles. Quite a few small biplanes use sweepback on the upper wing only in order to bring its MAC into proper relationship to that of the lower wing and the CG, while maintaining a far-forward location for the center section so that it will be ahead of the pilot for visibility and access reasons. Since there isn't room for him to squeeze in under it, it would seem logical to move him aft a bit, but this is a structural problem calling for moving the engine forward to compensate for the weight shift, etc.

The Proper Airfoil

Selecting the proper airfoil is one problem that amateur designers really suffer over, and so needlessly. Practically any of the common airfoils with a thickness of 10 to 18% of the chord will do. Tapered wings will tend toward 15-18% thickness at the root with 10-12% at the tips, while most straight wings will average 12-15%, with 15% favored for those using trailing edge flaps. The relatively new "Laminar Flow" sections look great on paper and promise wonderful performance advantages, but in most cases they are just another average airfoil when they get on the airplane. This is because amateur builders using practical types of home construction cannot work to the exacting tolerances that are required to make these sections pay off.

Craftsmanship, tooling, and structure that can hold tolerances to .001″ under load are way out of the homebuilt league. There are many other design parameters that have more effect on performance than the airfoil. One of the current high-performance German sailplanes with a glide ratio of 34:1, which is right up near the top of the scale, achieves this performance with the old reliable Clark Y airfoil, developed in 1922, coupled with an AR of 20.

The most popular airfoils for homebuilts are the Clark Y, the very similar NACA 4412, the NACA 23012, and the laminar family best represented by the NACA 63A415. Geometric ordinates and aerodynamic characteristics for these can be found in aeronautical engineering books and NACA (National Advisory Committee for Aeronautics, now National Aeronautics and Space Administration, or NASA) reports at Public Libraries and in various EAA publications.

Ailerons

These are the moveable surfaces located at the extremity of

the wing trailing edge and control the plane about the roll, or longitudinal axis. On the average, they are something under one-quarter the chord of the wing and between one-quarter and one-half the span, which works out to between 5 and 10 percent of the wing area. In general, long-span wings with higher rolling inertia require more aileron area.

There are many several different forms of aileron, and various ways of attaching them. By far the simplest is the plain aileron with a piano-type hinge at the top surface as shown by (A) in the following diagram. With the negligible gap at the hinge, there is no need to seal the gap against aerodynamic leakage. A variation is plain type (B) with the hinge on the spar centerline. For maximum effectiveness, this requires a fabric seal that crosses the hinge line.

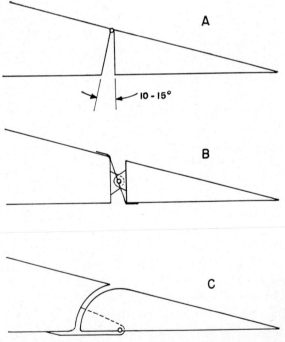

Both of these types have a slight aerodynamic disadvantage when rigged for equal travel up and down. This is known as adverse yaw — the aileron on the outside of the turn initially has more drag than the one on the inside, tending to swing the nose outward at the start of the turn unless compensated for by application of rudder at the proper time. Elimination of this characteristic is of paramount importance to the manufacturers of winged automobiles, where ultra-easy flight characteristics are an important sales feature, but it hardly bothers the sport pilot in an open-air flying machine. An easy way to eliminate it, however, is to use a control system hookup that will give differential aileron movement, about twice as much up movement (on the inside of the turn, where it produces more drag) than on the outside. Some big-aileron sailplanes go as high as 3:1.

One type of aileron that eliminates the yaw problem altogether and permits smooth turns without use of rudder is the "Frieze" type shown in (C). Because of the projected hinges and aileron area ahead of it, the leading edge of the aileron projects below the surface of the wing when the rest of the surface is deflected upward, adding considerable drag to the inside of the turn. However, the extra structural problems of projected hinges, cutouts, and the fairing structure are hardly worth the slight improvement in "Baby Buggy" flying characteristics.

Very few homebuilts have a need for mass-balanced ailerons. Because of the small size and light weight of the surfaces, mass balancing is useful only in counteracting aerodynamic flutter, which is not a problem at the airspeeds achieved by the average homebuilt. Aerodynamic balancing of plain ailerons is not necessary, although Frieze types can be considered to be aerodynamically balanced. "Drooping" of ailerons during landing to augment flaps and help slow the ship down is used only slightly on some military and cargo planes, and not at all on homebuilts. The effectiveness of the aileron decreases as it takes on the flap function, and can lead to a dangerous situation as its effectiveness is at a minimum anyhow during the slow speed and high angle of attack of the landing approach.

Flaps

Flaps, located on the inboard trailing edge of the wing, are used to provide additional lift during landing and takeoff. In this capacity they are simply a crutch used to partially overcome performance deficiencies resulting from heavy loads, inadequate wing area, etc. Since homebuilts do not carry heavy commercial payloads or fuel for extremely long ranges, they have little need for flaps. The handicaps of a 5 m.p.h. higher landing speed or a slightly longer takeoff run are easier to live with than the structural, mechanical, and aerodynamic trim problems that result from the use of flaps. Because of inefficiency resulting from small sizes, the effectiveness of the simple hinged flap on typical homebuilts is almost negligible.

Tail Surfaces

The rudder was named after the corresponding surface on a boat by the Wright brothers, and while it does not steer the ship, we have been stuck with the term ever since. It merely works in coordination with the ailerons to make trim corrections about the vertical, or yaw, axis. Most airplanes divide the vertical tail area, as determined by the formula (3) on the chart, between the movable rudder and the fixed vertical fin in a ratio from 50:50 to 60:40 either way.

Aerodynamic balance area, as used on many production planes and shown on the chart, is of no advantage on small and relatively slow speed planes (130 m.p.h. and under). Swept-back vertical tails on light airplanes are principally a matter of styling. Slight aerodynamic gains can be claimed from the fact that the sweep adds slightly to the effective vertical tail movement arm and that the inclined hinge line produces the equivalent of a touch of "up" elevator when the rudder is deflected during turn entry.

The horizontal tail surfaces (4) consisting of fixed stabilizer and movable elevator on most types, and again divided 50:50 or 60:40 in favor of the stabilizer, are usually located about even with the vertical tail. In recent years some homebuilts have followed the trend of the factory types in using one-piece "Flying tail" horizontal surfaces where the whole surface pivots at the fuselage.

These had been used on some early WW I fighters like the German Fokker "Eindecker" in the interest of lightness and simplicity, and on many famous gliders up to the late 1930's, but the major disadvantage was an almost total lack of the essential "feel" for the pilot. As a result, he could easily overcontrol at high speeds and overstress the ship.

This problem was eliminated after WW II by development of an "anti-balance" tab at the trailing edge that moved in the direction of surface displacement to resist the movement with aerodynamic forces and give the pilot the same feel proportional to actual air load that was produced by the standard two-piece arrangement. The aerodynamic advantage of the flying tail permits about a 15% reduction of size, with corresponding benefits in reduced weight and drag. The major problem facing the homebuilder is that of achieving a hinge loose enough to permit free movement and tight enough to prevent flutter. This is necessary because the hinges cover only a very small area and are highly loaded, transmitting the entire tail load to the fuselage. Horizontal tail surfaces divided at the centerline and braced on each side by struts or wires are the lightest and easiest to make.

While all production airplanes are equipped with adjustable stabilizers or elevator tabs to permit trimming the airplane longitudinally from the cockpit while in flight, it is about a 50:50 proposition on homebuilts. If the plane is to be used for extensive cross-country flying, where trim changes considerably as fuel is consumed, or a simple change of pilot weight affects the trim, a trim system should be used. This can be a jackscrew to move the whole stabilizer or a moveable tab on one elevator.

A prime requirement is that it be aerodynamically and statistically balanced, or else non-reversible, so that a sudden gust or other acceleration of the tail will not displace the tab. Tail surface hinges are located on the centerlines of the spars in the manner of (B) on the aileron diagram. On steel-tube structures, each hinge is made from three pieces of tubing, two on one surface and one on the mate, to hold a clevis pin or bolt. Wood and sheet metal structures use bracket-type hinges bolted or riveted on.

The Tail Moment Arm

The distance between the tail, especially the horizontal tail, and the CG has a significant effect on the controllability of the airplane. This is item (2) on the chart, and is generally called the tail moment arm. Ships with extremely short tails tend to be very touchy in the matter of fore-and-aft trim and are consequently very critical on CG location. Surface movement has to be greater to achieve the same effectiveness as a smaller movement on a longer tail. Too long a tail adds inertia and structural weight.

Many amateurs wonder why the wing loading can't be alleviated to some extent by letting the horizontal tail carry part of the actual flight load, and point to the many "lifting tail" model airplanes as an example. This used to be done on airplanes but was outlawed for acceptable design by the FAA many years ago.

Only Fleet and Stearman biplanes, production versions of designs developed and approved in the late 1920's, can be found flying with lifting stabilizers today. All tail surfaces in use today are symmetrical, with a thickness of 5 to 9% of the chord.

Fuselage

This has little effect on the aerodynamics of the airplane except as a means of maintaining tail length and the distribution of the weights carried in or attached to it. In small sizes and low speeds, the cross-section shape is of no significance, although unusual width, as on a short side-by-side two-seater, can affect tail efficiency by interfering with the airflow to it.

Other than a "dirty" nose design and an excess of protrusions, the principal drag-producing feature of a fuselage is its point of intersection with the wing. A good intersection between a curved-section fuselage and a low wing with considerable dihedral is a work of art, and a poor one costs much in drag and lost lift. Take a good look at this area on equivalent airplanes, both production and homebuilt, when you tackle this problem.

One of the major problems in fuselage layout is working out a satisfactory arrangement of all the things that have to go into it. All too often the pilot is the one that suffers from the compromises. It should be the other way around, for pilot comfort, good visibility, and easy access are very essential to the usefulness of the ship as a sportplane.

The engine location is a multiple compromise between balancing the pilot and other loads, the position of the propeller ahead of the wing leading edge per item (9) of the chart, and a vertical thrust line location, partly determined by the landing gear configuration and partly by the required propeller diameter for the particular engine, that will assure adequate ground clearance per item (11).

Open vs. closed cockpit configuration is a matter of pilot preference, not performance improvement. Some pilots like them open, and the visibility is definitely better. A compromise between fun and comfort is the sliding canopy between the windshield and a built-up turtledeck between the cockpit and the tail. This improves the streamlining but in most cases cuts down drastically on visibility to the rear. Some designs eliminate the turtledeck by using a blown or moulded "Bubble" canopy over the cockpit. These are especially popular on sailplanes and midget racers.

Cockpit Location

Cockpit location is a very important factor in the selection of a design for a reduced-scale replica. If the original is a biplane with a cockpit in the desirable position on or near the CG, the pilot might not be able to get into it because of the reduced wing/fuselage gap. If the cockpit is aft of the upper wing, an impossible balance situation may result because the weight of the reduced replica will be half that of the original or less, while the pilot's weight stays the same and is not adequately balanced by the smaller engine, which usually turns out to be $1/4$ to $1/3$ the weight of the original.

Fuselage width is a problem in scale replicas, too. The fuselage of a Curtiss "Hawk" fighter, often considered a desirable replica item because it is such a famous and good looking airplane and has a 31'6" wingspan which would reduce to 23 feet, just right for a homebuilt in $3/4$-size, is only 24" wide. This is the width of the fuselage of the average single-seater or tandem two-seat homebuilt. A true scale $3/4$ replica would use a fuselage with the impossible (for a pilot) width of 18". A compromise with true proportion is therefore necessary.

Landing Gear

The majority of homebuilts use the old "conventional" landing gear configuration with two wheels in front and a small steerable wheel under the tail. Hardly any "Cockpit" designs use the tricycle gear, which is better suited to cabin types. Main gear location has a great influence on airplane take off and landing performance and on handling characteristics on the ground. Item (7) from the chart is wheel location as specified in the "Handbook of Instructions for Airplane Designers" published by the U. S. Air Force.

A wheel location too far forward has great anti-noseover tendencies but is "unstable" for a ground run because the CG is quite far behind the main wheels and will tend to turn the ship around if it wobbles a bit during the landing roll. Moving the wheels farther aft tends to correct this end-swapping characteristic, but increases the tendency to nose over as the brakes are applied.

The height of the conventional landing gear is also quite important. For best takeoff and landing, the wing, when the airplane is at rest, should have an angle relative to the ground

(8) just equal to that at which it generates 90% of its maximum lift. This is important to takeoff, but even more important to a slow landing. The plane cannot fly at its slowest speed near the ground if the tailwheel hits the ground and decreases the angle of attack of the wing. Planes with relatively flat ground angles must be landed "hot," usually under power by the technique known as "wheel landing." While a high angle is good for slow landing, it is a handicap to forward visibility from an aft-located cockpit during taxiing, which can be a safety hazard.

Tricycle Landing Gear

Tricycle gear is directionally "stable" on the ground because the main wheels follow the CG and the ship tends to straighten itself out if it yaws a bit on the landing roll. Since the tricycle ship sits level on the ground, visibility over the nose, especially from a cabin, is greatly improved. The principal handicaps to tricycle gear on small airplanes are the extra weight and drag, and on all sizes there is the need for beefed-up fuselage structure where the nosewheel strut ties in and a shimmy-free steering mechanism.

The Cockpit

Unfortunately, this usually turns out to be the most-neglected feature of the entire design, when actually it is one of the most important. Not only does the pilot have to be able to get into it, be reasonably comfortable, and see out of it, he has to be able to operate a number of controls, scan and interpret instrument readings, and sometimes handle a communications system. His ability to do any of these things is not just a matter of luck—it has to be intelligently planned.

The problem begins with access and comfort, (and don't forget that in cold weather you'll be dressed in gear and possibly wearing a parachute.) To accomodate pilots of varying size, either the rudder pedals should be adjustable or the seat, preferably both. On open cockpit types, the windshield should give really good protection. Many sporty-looking little shields actually contribute to pilot discomfort by stirring up the slipstream so that it batters him from both sides as well as front and back. Curved glass or plastic types that are large enough to have a sliding canopy, or hatch, mated to them are usually large enough to give good protection when the canopy is slid back.

Rudder pedals should not be too close to the pilot's seat. This forces him to raise his knees, which may hit the instrument panel, and the position is tiring at best. Shoulder and elbow room is important, and the instrument panel should not be too close to the seat. While some designs use narrower ones, a 24-inch fuselage width at the cockpit of a single-seater will take most pilots comfortably. Distance from the surface of the seat to the top of cabin or canopy should be a minimum of three feet.

Layout of the controls relative to the seat is very important. If the control stick is in an awkward position, the portion above the fitting can be bent forward or aft for pilot convenience. If a parachute is to be used at all, the seat should be worked out with the specific type in mind, whether seat or back-pack.

The sequence of control operation during normal flying should determine control layout; the pilot should not have to switch hands from the control stick, or operate something on the right side of the cockpit with his left hand. Most single and tandem-seat airplanes today mount the throttle on the left side of the cockpit, along with such other engine controls as switches, carburetor heat, mixture, etc, so that sequential operations can be performed with the same hand. Side-by-side two seaters have the throttle between the pilot and copilot, so the one in the left develops right-handed throttle habits while the man on the right develops his left as he would in the tandem types. A pilot's previous flying habits may determine his choice of throttle location in his own homebuilt.

Instruments should be grouped generally according to function —all the engine instruments in one part of the panel, the flight group in another where they can be scanned as a unit, and the navigation equipment elsewhere. Frequently, the only navigation instrument—a lone magnetic compass—is mounted clear away from the panel, up at the top of the windshield or under the center section of a biplane. This gets it away from heavy metal items and other objects that cause interference. Magnetic tachometers, especially, should be kept away from the compass. Consideration should also be given to mounting of the instruments and access to them for servicing. When possible, make the instrument panel as a unit, or at least a section of it, easily removable.

Engine-Propeller Combinations

Except in rare cases for special installation, the home builders (and even the commercial airframe manufacturers) do not design and build their own propellers. These are included in the "stock hardware" items that are purchased ready-made. The prop has a great effect on the performance of the plane, far more than the average pilot/designer realizes, and coming up with exactly the right one for a particular airplane/engine combination usually turns out to be something like buying shoes—if you don't know the exact size ahead of time, you keep trying until you find one that is just right.

A little diligent research will pay off here, because you can't try several at the store before buying the right one. The answer to the question won't be in the books—go to the airports and look at the airplanes. Find out what equivalent airplanes with your engine are using, and start there. If you already own a one-piece metal prop, any approved repair shop can cut it down or re-pitch it so that it is just right for your engine and airspeed.

It may come as a surprise to some pilots that more powerful engines do not necessarily swing bigger propellers. As an example, the 65-h.p. Continental turns a 74-inch diameter prop of 44-inch pitch at 2,150 r.p.m. for cruising. ("Red Line," or never-exceed speed for the A65 is 2,350 r.p.m.). When the same engine is modified to an A75 of 75 h.p. with a conversion kit at no increase in displacement from the original 170 cubic inches, the extra 10 h.p. comes mainly from running it 250 r.p.m. faster, cruising at 2,400 r.p.m. and raising the "Red Line" to 2,600. This is accomplished by *Reducing* both the diameter and pitch of the prop to 69 or 70 and 38 inches, respectively, for the same airplane. The C.85 Continental, on the other hand, a 188-cubic inch engine, swings a 71-inch prop of about 48 to 51-inch pitch at 2,350 for cruise and 2,575 "Red Line."

Propeller Diameter

Prop diameter is not an arbitrary thing that can be altered at the whim of the builder. Damaging vibration modes can be set up by using the wrong diameter, especially in metal propellers, for a given engine. Sometimes there are critical r.p.m. settings within the normal range of operations that will set up vibration. In production planes, the entire range is tested thoroughly with very precise instrumentation, and any critical areas that are found are placarded to warn the pilot to avoid them. The amateur can only watch for an indication of vibration areas within his operating range and take care to avoid them or pass through them as quickly as possible.

Diameter, since it affects the area of the propeller, also has an effect on the r.p.m. that the engine can deliver. In general, the smaller the prop, the faster it can be turned to get more power out of the engine. Large props cannot be turned as fast because the speed of the tips, a function of blade length as well as r.p.m., must be kept well below the speed of sound. In big planes this is taken care of by gearing a high-speed engine down to a relatively low prop r.p.m. While increased crankshaft speed delivers more power from an engine of a given displacement, the r.p.m. increase is not all pure gain. The efficiency of the prop drops off drastically as the r.p.m.'s increase. The modern Continental 85 delivers its maxiumum output at 2,575, while the 80-h.p. Le Rhone rotary of 1915-18 topped out at 1,250 with a wide-blade eight-foot prop.

On the other hand, the pilots of the midget racers, all of which use the C.85, sometimes turn 60-inch toothpick props at 3,600 r.p.m. in the belief that the H.P. gain from the extra 1,025 r.p.m. will more than offset the decrease in blade efficiency. The McCullough O-100 (100 cubic inch) target drone engine delivers its 72 h.p. through a tiny prop that turns 6,000 r.p.m. but has nowhere near the thrust of an A.65 turning 2,350.

Getting the Right Pitch

The "pitch" of a propeller is the distance that it moves forward for each complete revolution under optimum conditions. The term is only generally understood and widely misued. A prop with high blade angle is said to have high pitch, but since pitch is a function of the forward movement of the airplane, it can be seen that the two terms are not exactly synonymous. For a given airplane and engine, the pitch of the prop (actually the blade angle that controls the r.p.m. and consequently the pitch) must be set for the most-desired condition.

If the plane is to be used primarily for cross-country work where maximum cruising speed is desired, the pitch should be increased. Such propeller is referred to as a "cruising prop." Getting this increased performance at the high end of the scale costs, and sometimes quite critically, on the lower end. Because of the high blade angle suited to high-speed cruise, the blade angle is too high relative to the air at take off and climb speed and the added drag keeps the engine from delivering its maximum power.

The opposite procedure is to set the pitch to a lower angle so the engine will turn up better at the lower airspeed. This is called a "climb prop," or sometimes a "seaplane prop," since the drag of the floats alone cuts the airspeed down to a point where the blade angle should be decreased to be compatible. When using climb props, the pilot must take care to keep the engine from overspeeding in level flight.

Since one-piece metal props, with their thinner blades, are so much more efficient than equivalent wooden ones, the homebuilder can adopt a compromise setting that will give him as good a performance on each side of the median as he could get with either a climb or cruising prop made of wood. For the metal props used on 65-90-h.p. lightplanes and homebuilts, a change in pitch of three inches is good for an engine r.p.m. change of 100 in level fllight. Controllable-pitch props that can be adjusted in flight are available, but they are quite expensive, and in the sizes used on engines up to 150-175 h.p. do not do a good job because of the loss of efficiency from thick blade shanks and the large hub. Even the industry doesn't bother with controllable-pitch props for planes under 175 h.p.

Miscellaneous

Entirely aside from primary and detail design considerations, careful attention must be paid to such thingh as access and maintenance. It is in this area that previous experience in servicing and maintaining aircraft pays off. Many high-time pilots who do no work on the planes they fly have no concept of these problems and so do not consider them when designing a homebuilt. One of the most common errors is to design "blind" installations where it is impossible to gain access to the back side in order to tighten the bolts, etc. A similar situation results from lack of thought given to the accessibility of a part that may have to be removed for maintenance or repair. Such items should be removable directly without having to remove several other items first in order to get to them. Items to be serviced in the plane, such as hydraulic brake master cylinders, should be accessible. While they may be above the floor, they are apt to be at the far forward end of the cockpit and under the fuel tank, impossible for anyone but a midget acrobat to service. An access panel in the side of the fuselage right opposite each unit will eliminate the problem.

Adequate drainage is another area frequently neglected by amateur designers. Trapped moisture—either water or gasoline and oil—standing in a wing or fuselage can make any airplane, wood or metal, unairworthy in a short time. The structure should be carefully analyzed for built-in moisture traps in areas not normally cleared by the use of standard aircraft drain grommets.

Pontoon seaplanes are rarities among homebuilts for reasons given on page 26. Author's "Fly Baby," normally a landplane, flies with antique Edo model 990 pontoons built in 1932. Speed decreased both by added drag of floats and decrease in propeller pitch to meet seaplane requirements. This size float gave marginal flotation, and larger model 1070 is better suited to weight and power of plane in this size/weight range. (Photo by James L. Slauson, 1963)

MONOPLANES — SINGLE SEATERS

Make and Model	Powerplant	Wing Span	Length	Wing Area	Empty Weight	Gross Weight	High Speed
Bowers Fly Baby I	85 Continental	28' 0"	19' 0"	120 sq. ft.	605 lbs.	924 lbs.	120 mph
Corben Baby Ace D	65 Continental	26' 5"	17' 8¾"	112 sq. ft.	575 lbs.	950 lbs.	110 mph
Druine Turbulent	30 Ardem (VW)	21' 5"	17' 4½"	80 sq. ft.	341 lbs.	606 lbs.	87 mph
Hueberger Doodle Bug	90 Continental	18' 0"	16' 6"	68 sq. ft.	616 lbs.	956 lbs.	202 mph
Jodel D-9	25 VW	22'11"	17'10½"	96.8 sq. ft.	356 lbs.	598 lbs.	93 mph
Long Midget Mustang	85 Continental	18' 6"	16' 0"	68 sq. ft.	575 lbs.	875 lbs.	190 mph
Loving Love	85 Continental	20' 0"	17' 2"	66 sq. ft.	631 lbs.	839 lbs.	215 mph
Smith Termite	65 Lycoming	23' 6"	15' 9"		726 lbs.	976 lbs.	90 mph
Solvay-Stark Skyhopper	65 Continental	25' 0"	18'10"	100 sq. ft.	650 lbs.	950 lbs.	130 mph(?)
Stits Playboy SA-3	85 Continental	22' 2"	17' 4"	96 sq. ft.	600 lbs.	902 lbs.	145 mph

BIPLANES — SINGLE SEATERS

Make and Model	Powerplant	Wing Span	Length	Wing Area	Empty Weight	Gross Weight	High Speed
EAA Biplane	65 Continental	20' 0"	17' 0"				105 mph
Knight Twister	90 Continental	15' 0"	14' 0"	60 sq. ft.	535 lbs.	960 lbs.	160 mph
Meyers Little Toot	90 Continental	19' 0"	16' 6"	123 sq. ft.	914 lbs.	1,230 lbs.	127 mph
Mong Sport MS-2	65 Continental	16'10"	14' 1"		540 lbs.	750 lbs.	115 mph
Smith Miniplane	100 Lycoming	17' 0"	15' 1"	100 sq. ft.	616 lbs.	1,000 lbs.	135 mph
Sorrell Special	65 Continental	22' 6"		100 sq. ft.	505 lbs.	740 lbs.	95 mph
Stolp-Adams Starduster	125 Lycoming	19' 0"	16' 6"	110 sq. ft.	700 lbs.	1,080 lbs.	147 mph

MONOPLANES — TWO SEATERS

Make and Model	Powerplant	Wing Span	Length	Wing Area	Empty Weight	Gross Weight	High Speed
Hueberger Sizzler	125 Lycoming	18' 4"	20' 7"	100 sq. ft.	900 lbs.	1,400 lbs.	175 mph
Jodel D-11	65 Continental	26'10"	20' 4"		594 lbs.	1,067 lbs.	106 mph
Nesmith Cougar	115 Lycoming	20' 6"	18'11"	82.5 sq. ft.	624 lbs.	1,216 lbs.	182 mph
Piel Emeraude	65 Continental	27' 3"	19' 9"	116 sq. ft.	583 lbs.	1,100 lbs.	112 mph
Pietenpol Air Camper	65 Franklin	28' 2"	17' 8"		630 lbs.	1,050 lbs.	105 mph
Stits Flutterbug	65 Continental	26' 0"	18' 0"	130 sq. ft.	575 lbs.	1,031 lbs.	100 mph
Stits Playboy SA-3B	150 Lycoming	24' 2"	17'10"		795 lbs.	1,450 lbs.	150 mph
Stits Skycoupe	85 Continental	25' 9"	17' 9"	120 sq. ft.	650 lbs.	1,175 lbs.	138 mph(?)
Trefethen Sportaire II	125 Lycoming	26' 4"	20' 6"				
Volmer Sportsman	85 Continental	36' 6"	23' 6"	175 sq. ft.	950 lbs.	1,450 lbs.	100 mph
Wittman Tailwind	85 Continental	20'11"	19' 3"	83.5 sq. ft.	700 lbs.	1,300 lbs.	170 mph

6. Construction Problems

Building an airplane is not a big job. It is a whole collection of little ones whose degree of difficulty, independent of the builder's skill, is influenced to a large degree by the available work area and conditions and by the tools and methods used. Other than the actual details of making airplane parts from drawings, the construction problems faced by the average homebuilder fall into three general areas: the working area and the tools involved, work practices, and personnel.

Working Area and Conditions

Some amateur-built airplanes have been built in surprisingly illogical places, which proves that formal shop facilities are not an absolute necessity. However, suitable space where the work can be left standing is desirable. An area equal in size to a standard one-car garage is just about the minimum that can honestly be called suitable for turning out a typical homebuilt airplane

There should be enough area to lay out a 5 x 14 foot wing panel on saw horses and still leave room to work around both sides and the ends, with additional space for a workbench and the standing tools. Fuselages average 14 to 15 feet long without the engine mount or landing gear, and are usually two feet wide with a few going to three and one-half feet. Without the landing gear attached, most homebuilt fuselages can be taken through a standard door if it is so located relative to the work that a straight approach can be made to it. This means that homebuilts can be built in many rooms of a regular house and then be removed without knocking out a wall in the classic "boat-in-the-basement" tradition. If space is tight, consideration must be given to a separate area for the unused raw material and for major components of the plane as they are finished.

Since glueing with Weldwood and all doping must be done at temperatures above 70 degrees Fahrenheit, heat control in the working area is essential. Other glues are available for lower temperatures, but with no heat control the doping may have to be deferred until warm weather.

Doping should not be done in a confined area without adequate forced ventilation, so unless a suitably-equipped shop is available the work should be done outdoors or taken to another shop.

Nothing will kill family approval of a homebuilt airplane project faster than a house full of dope fumes.

Tools

As with the working area, airplanes can be built with a bare minimum of tools, but the job is greatly simplified by having a proper selection for the various jobs. Items marked with an asterick (*) in the following list are considered absolutely essentiel to the building of any airplane regardless of the type of construction. If these are not on hand in the home shop (and it is not logical to buy some expensive items for a single five-minute job) the material will have to be taken elsewhere for processing or the needed tool must be borrowed or rented. Aside from a suitable power saw and a drill press, the most essential single tool will be the electric hand drill and an extension cord and light. As the work progresses, much more work will be done on and inside the airplane than on the bench.

Experienced wood workers, familiar only with boat and furniture construction, are often surprised at how few of the traditional woodworking power tools, shapers, joiners, etc., need be used in small wooden aircraft construction. Mortised or dovetail joints are not used in aircraft, and long strips can be taken down to very small dimensions, like $1/4''$ square, on a table or radial arm saw fitted with a "planner" blade. On the other hand, a surprising amount of woodwork is involved in the construction of an all-metal airplane in the form of jigging, pattern blocks, etc.

SAWS
- Table or radial arm saw*
- Coping saw or jigsaw
- Bandsaw or Bayonet saw with stand*
- Hacksaw*
- Small strongback saw or hand saw

DRILLS
- Bench drill or drill press*
- Electric hand drill*
- Drill bits to max. dia. $3/8''$ (with $1/4''$ shank for hand drill)*

FILES
- Suitable flat, rounded, and rat-tail files for wood and metal finishing*
- Coarse wood rasp

Rotary file for electric drill
Rotary rasp for electric drill

C-CLAMPS (Suitable Sizes and Quantities)

VISE (steel)*
TACK HAMMER
BENCH SANDING DISC
BENCH GRINDING WHEEL
BLOCK PLANES
TIN SNIPS*
DRAW KNIFE OR SPOKE-SHAVE
3' CARPENTERS SQUARE
6' STEEL MEASURING TAPE*
STANDARD AND PHILLIPS SCREW DRIVERS*
DIAGONAL CUTTERS ("DIKES")*
PLIERS*
PAINT SPRAY OUTFIT
COMBINATION SQUARE*
SOLDERING IRON
PINKING SHEARS

WELDING RIG
SAW HORSES (2 minimum)*
GLUE AND DOPE BRUSHES*
PAPER CUPS (unwaxed, for glue and varnish)*
COLD CHISEL OR WIRE CUTTERS*
NICOPRESS SQUEEZERS
BOX OR OPEN-END WRENCHES TO $3/4''$ *
SHARP WOODWORKING KNIFE
SAILMAKER'S NEEDLES
RIVET GUN
BUCKING BARS
SHEET METAL BRAKE
SHEET METAL FASTENERS ("CLECOS")
SANDPAPER AND EMERY CLOTH*

Work Practices

The home-built airplane project succeeds or fails in this area. The cost of the project is the very least of the reasons for giving it up. Storage and working area problems are a major cause of quitting, but the principal one is lack of progress and subsequent disillusionment resulting from poor work practices and lack of planning. The work appears to go fast in the early stages when major components are assembled, but seems to stand still in later stages when all the little things go inside.

As previously stated, building an airplane is a collection of many little jobs. Try to do them one at a time instead of having several going at once. Make a list of things to do in a relatively logical sequence, or based on the materials that are available at the time. A job list is very handy thing to have on those occasions when a competent friend drops into the shop and offers to help out. You might not be able to think of anything on the spur of the moment, but the list may reveal something suitable.

Be systematic in keeping track of tools and materials. Much time is lost in merely looking for some small tool that you need to use only a minute or so. Similarly, always buy a bit more material than that called for on the parts list, at least in the nuts-and-bolts department and in smaller sizes of raw material. It seems that one always drops the last bolt of a certain size through the crack in the floor on Saturday evening just after the store closes, and no other work can be done until the item using that particular bolt is installed. Running short of material on the eve of a holiday produces the same situation. The cost of extra material is practically negligible, but the few dollars spent can buy a lot of valuable time that might otherwise be lost.

While circumstances will dictate different procedures for different people because of equipment, availability of materials, etc., these time-saving suggestions can be followed by almost everyone:

Cut as Many Pieces of a Size as Possible at One Time. Much time is wasted in resetting the tools (power saw, drill, etc.) when pieces are cut singly or a few at a time on an "as-needed" basis. Consideration must be given, however, to the stipulation in F.A.A. Manual 18 that wood surfaces for gluing should not be exposed for more than 24 hours prior to gluing.

Work from Big Pieces Down to Small. For trimming parts to fit

A simple working surface. Rex Richards, of Seattle, assembles the hull of a wooden flying boat on a work table made up principally from two heavy sheets of standard commercial 4-foot by 8-foot fir plywood. (Photo by Peter M. Bowers, 1959)

it is almost inevitable that an occasional piece will be cut a fraction of an inch too short. If the work sequence is such that the piece can be trimmed to a smaller size for another installation, waste of material and time is avoided.

Mix Glue With Specific Jobs in Mind. Much expensive glue is wasted by mixing too much for a particular job. The "pot life" of Weldwood glue is only four hours, so it can't be saved for tomorrow. If quite a few items are to be glued over a fairly long continuous period, like an afternoon of installing wing rib corner blocks, plan on mixing several small batches during that time. Small batches are easier to mix and there is no question of approaching the pot life limit as the job goes on. Similarly, take precautions against running out of glue in the middle of a big and fast job such as laminating wing tip bows. Mix several batches rather than one big one, or have a helper mixing new ones as you use the first.

The best applicator for glue is a 1/2" to 3/8" paint brush. If the brush is washed out in hot water before the glue sets, one brush can last for months. Glue is cheaper in 5-pound cans, but constant opening of the can to take out small quantities ages it rapidly. It is best to buy it in small cans.

Plan Varnishing So as Not to Block Other Work. Try to save varnishing that will hold up other work in a particular area for the end of the work period so that it can dry overnight or between sessions. When wet varnish is on some parts while others are being worked on, be sure that shavings and chips don't fall on the varnish. Remember that dust from saws and grinders can float all over the shop and settle on a wet varnish job clear across the room from the tool.

Don't open the varnish can for small jobs. Poke two nail holes on opposite sides of the lid and seal them with pieces of masking tape. Pull the tape and pour varnish into a paper cup for mixing with turpentine for small jobs. The can gets messy if varnish is poured over the lip, and after several openings for pouring small quantities, the varnish begins to thicken and scum over. Although cheaper by the gallon, it is best to buy varnish in quart cans. Be careful not to use waxed paper cups for varnish, glue, or dope. "Hot cups" are fine.

Use Systematic Work Habits. Try to plan the work for several days or individual jobs ahead so as to have all the necessary

Only bachelors can get away with this. Airline pilot Dave Gauthier assembles the wooden wing of his Gauthier "Sport" in his bedroom. (Photo by Peter M. Bowers, 1959)

material on hand, and organize the most efficient sequence for doing things. Much time can be lost by wondering "what to do next?" and then figuring out how to go about it. Try to work on related jobs in sequence so that parts for several can be cut at one time. Try to set up specified times for working, with an ideal objective of being able to get some little thing, even if it's only removing the clamps from yesterday's work, done every day.

Personnel Problems

Building an airplane is much more than just a technical problem. The people that are involved, actively or passively, sometimes provide the really serious problems. One of the major roadblocks to completion of any home workshop project is objection by authoritative members of the family if legitimate family obligations and relationships are neglected for the project. This is a political matter beyond the scope of this technical document, but is nevertheless a major item for consideration. Other than the family situation, there are three major human causes of wasted time in construction projects.

The first is the eager friend who is anxious to be helpful but doesn't know anything about building airplanes or even handling tools. By the time you show him how, check his work, and generally do it over, you could have done it several times yourself in addition to the job you were working on. The exact and highly

Elaborate jigging is not needed to assemble an airplane. The two wooden sides of a Bowers "Fly Baby" fuselage are temporarily joined by notched 2 x 4's used as spacer blocks while the permanent cross pieces are installed. (Photo by Peter M. Bowers, 1960)

desirable opposite of this type, and unfortunately very rare, is the experienced person who can be handed a job and forgotten for a while as he gets it done with no fuss.

The second time-killer, more often plural than singular, is the curious and friendly type who drops around from time to time "to see how you are doing" and brings a friend along who has to have the whole project explained in detail from the very beginning. No work can be done at all during most of these visits, and the visitors are very seldom inclined or even qualified to help. A sub-category of this type is the one with whom a little knowledge is a dangerous thing, and who is always trying to improve your design to death by suggesting all sorts of things from little refinements to major rearrangements that will be made with YOUR time, money, and materials.

One unforeseen by-product of both categories of this second type is the added expense to the overall job resulting from the amount of your groceries, coffee, beer, etc., that they consume while sitting around keeping you from working.

The third major thief of your working time is yourself. As the plane begins to go together it is entirely too easy to gaze dreamily at it by the hour, admiring your own handiwork and engaging in all sorts of flights of fancy while sitting in the cockpit of an unfinished fuselage perched on a pair of sawhorses. Even if you don't feel particularly ambitious when you go out to the shop, try to make *some* tangible progress. Don't goof off for one whole work period by kidding yourself with the thought that you'll really bear down on it "tomorrow."

Overdoing the improvements can be a personal matter, too, although in this case it usually stems from improving skill as the job progresses. There may be such a difference between the first few ribs you built and the last that you will want to scrap the early ones and do them over. Your own standards and time/cost considerations will be your only guides here.

7. Testing

While the term "testing" is associated by most people only with the completed airplane — generally the first flight and subsequent proving and adjustment flights — it is actually a continuous process that begins shortly after the airplane starts to go together. Hand-holding a part in position to see that it fits before attaching it permanently is just as much a test operation as running the engine for the first time.

A little extra attention paid to details during the construction stages can save many hours of disassembly and rework when the plane is supposed to be ready to go. As various components are built, they should be checked against the larger assembly that they attach to while there is still time to make adjustments before firming them up. A ¼-inch discrepancy between wing fittings and the points at which they attach to the fuselage can be very embarrassing and can delay the flight phase of the test program for quite a while. All moving parts should be tested for full travel without interference during the early installation phases. Check for overtravel as well as for insufficient travel.

Testing of the finished airplane begins with the powerplant, and can be done before the wings are attached if the fuel tank is in the fuselage. Anyone with automotive experience can check the continuity of the wiring. If you don't have this experience, get an aircraft engine mechanic to help. One significant difference between airplane and automobile engines is that the airplanes use magneto instead of battery ignition, and dual ignition at that. The fact that a magneto switch is "On" when the contacts are open, which would normally be "Off" for ordinary switches, is something that should be checked out carefully.

Another very important test is fuel flow. The engine manufacturers provide data in their specification sheets for the required fuel head and rate of flow through the lines for the simple gravity-flow systems used by most of the production lightplanes. Because of the smaller dimensions of homebuilts, the fuel head problem (height of the tank outlet above the carburetor) can get quite critical.

While an installation may be satisfactory for level flight, flow may be inadequate in climb attitude because the location of the tank aft of the engine can put the outlet below the carburetor when the tail is down, as in a climb, or just in the normal attitude with all the wheels on the ground. An easy way to check for fuel flow in steep climbs is to dig a hole for the tailwheel so that the nose is pointing upward about 30 degrees instead of the usual 15-18 degrees.

Tail-In-A-Hole

Testing of the engine in the plane prior to first flight is a mandatory FAA requirement: It must be run for at least an hour at speeds up to full throttle. This is supposed to reveal deficiencies in wiring, fuel flow, cooling, etc., so that they can be corrected prior to flight. While the regulations do not go into sufficient detail to specify a full-throttle run in climbing attitude, it is an excellent idea to do this by the tailwheel-in-a-hole technique. The initial climb from takeoff is one of the most critical points in the first real flight and is no place to have the engine quit from fuel starvation.

While the engine and propeller may have been installed in the shop and resulted in adequate ground clearance for the prop with the plane in level position, the clearance may be inadequate when the entire plane is assembled and all the added weight, including pilot and a full tank of gas, deflects the shock absorbers. Nine inches is the minimum for a fully loaded plane in level (takeoff-run) attitude.

Care should be taken during the engine tests (and during all engine runups prior to regular flight) to do the running over a paved surface whenever possible. The top vortices of a fast-revving propeller can suck up gravel and small stones and not only nick the propeller tips but damage the belly and tail surfaces of the plane itself.

Before the plane is taxied or flown, two essential checks should be run. The first is for proper weight and balance. If the plane is built from published plans and the builder hasn't made too many changes or doubled the strength of everything, the Center of Gravity (CG) should come quite close to the allowable range specified by the designer. If the plane is an original design for which the limits have not been established, anything between 20 and 30% of the mean aerodynamic chord of the wing is safe for a start. The maximum range that any homebuilt can tolerate is 15-35%. Procedures for weighing both conventional and tricycle-landing-gear airplanes are presented in FAA Manual 18.

Few amateur builders can be expected to have the right type of scales, (three of them at that), available for the normal weighing procedure. A single bathroom scale will do if the various components are weighed as the plane is built. Build a little jig or cradle that will set on the scale platform and hold odd-shaped parts, or just stand on the scales while holding the part then weigh yourself alone. It is possible to weigh a completed airplane with only two bathroom scales (normal capacity 250 lbs. each) by weighing each wheel separately on a wood bridge spanning both scales. It is important that the two other wheels be blocked up so that the airplane is in the proper level attitudes when the scale platforms are deflected.

A simple method of locating the CG, entirely independent of determining weight, is to suspend the whole plane from a chain hoist by a bridle assembly fastened to suitable strong points. When the plane hangs level, the CG will be directly under the point where the bridle hangs from the hoist hook. If the CG checks out fairly close to either limit, make a mathematical check per the procedures of Manual 18 to determine the effects of full vs. light fuel load, baggage, and the weights of various pilots that can be expected to fly the machine. These changes must not be allowed to run the CG beyond the limits.

The other essential preflight check is proper alignment and rig of the wings and tail. If the plane has been carefully built from proven plans, this should not be a problem, but a careful check should be made and any misalignment corrected before attempting to fly.

For a new design, where the optimum settings are not known but must be determined by flight test, a check should still be made for "squareness" of the rig — determining that both wings have the same amount of dihedral and the same angle of incidence relative to the fuselage. These are hard to measure, and can be checked best by the "eyeball" method — sighting the airplane from a distance head-on and checking the symmetry visually. Dihedral can be measured at the fuselage on some types by use of an adjustable protractor. Accuracy of lateral rig can be determined by running a wire from the same point on each wingtip to the top of the rudder post (chalklines or strings are unsatisfactory — they stretch).

Testing The Airframe

With everything satisfactory in the weight and powerplane departments, the airframe itself is ready for testing. This begins with the normal preflight check of the controls for free and proper movement (you'd be surprised at how many people install their control hookup backward and don't discover the fact until the plane is at the airport). Next comes taxi-testing to check alignment of the wheels, proper adjustment of the brakes, and adequacy of the tail wheel steering. Start this by letting the airplane taxi slowly with the control stick FULL BACK, and go easy on the first brake applications. If the plane is a bit nose-heavy, or the landing wheels are a little too far aft of the CG, a combination of a little too much speed and heavy braking can easily nose it over. If the plane demonstrates any nose-over tendencies during this phase of the testing, make corrections before going any farther.

With the slow-speed taxi testing satisfactorily completed, the final phase before actual flight is high-speed taxiing. The longer the runway available for this testing the better. The pilot can learn much more about the ship if he can hold a steady course for a while instead of having to stop almost immediately after he starts. This consists of fast dashes down the runway with the tail in the air at not quite enough speed to get the machine airborne, and provides a check on proper alignment of the wings, the effectiveness of control surfaces, and the trim of the machine.

As the tail is brought down to bring the ship to a stop, it also provides a final check of the directional stability of the landing gear and the proper relationship of wheels and CG. Planes with very short tails, or with the main wheels too far forward of the CG, tend to be directionally unstable at high ground speeds. So if the ship demonstrates these tendencies, and they are not easily correctable, the pilot should do a lot of taxi-testing to get thoroughly used to the characteristic so that he can handle them. The fast runs will also give an indication as to whether or not the airspeed indicating system is working.

Frequently, high-speed taxi-testing gets to the point where the ship gets light on the wheels and just happens to "bounce" high enough to get a considerable amount of daylight under the wheels. The line between legal high-speed taxiing prior to FAA approval for flight and actual free-flight conditions is very thin here. More than one pilot has built up too much speed under these conditions and "bounced" so high that he couldn't stop before hitting the fence and so had to fly the ship around the pattern. These things are awfully hard to explain to the FAA inspector and can easily sour any pleasant relationship that may have existed previously. If you plan extensive taxi-testing before getting the inspector's ok for flight, talk it over with him and tell him exactly what you intend to do and get his approval.

While taxi-testing can be done at any airport that will permit it, and the pilot involved does not have to be rated in the airplane or even licensed, FAA regulations state very specifically that the first flight test of an amateur-built airplane shall be made by a pilot holding at least a private license and a current rating in that category of aircraft. This means that if the airplane is single-engine land, the pilot must have an "airplane-single engine land" rating on a private license or better. If it's a rotorplane or glider, he must have the appropriate rating. See Chapter 8 for other legal considerations of testing.

While it is understandable that anyone having gone through all the work of building his own airplane would also want to have the satisfaction of making the initial flight in it, this is not always a wise procedure. If the builder is a relatively new pilot, experienced only in docile machines like Piper Cubs, he should get someone else to make the first tests on a hot little biplane or something that is unconventional. Pilot skill here can do a lot to overcome airplane deficiencies that would get a green pilot into serious trouble. In such cases, a little punctured pride is a lot less costly than a bent airplane.

In most cases, the FAA inspector will want to witness the first flight. This will probably mean that it will have to be during the week, within regular FAA working hours. A lot of homebuilders have found that this provides an unexpected advantage — practically everyone they know wants to be on hand for the first flight, and making it on a working day will keep many of them away and eliminate the need for the builder to contact everyone or offend someone by overlooking him. A crowd, with everyone eager to see the bird fly, is a serious psychological handicap to safe testing. Its very presence puts pressure on the pilot to go ahead with the flight in spite of little deficiencies that would ordinarily cause him to postpone the test in order to get everything perfect.

The First Flight

After approval for a full flight test and an airworthiness certificate have been obtained from FAA, the first actual flight should be a "slow" hop for the full length of the runway with a landing straight ahead. This is little more than a long extension of a "bounce" during high-speed taxiing. One of these may be enough, although several with increasing speed of takeoff can determine the effect of acceleration on the fuel flow. Because of the practically zero altitude, a parachute is of little use unless it affects the pilot's position in the plane for his forward vision. For the first flight out of the pattern, a 'chute is desirable but not mandatory.

Before making the first flight beyond the runway, fill the gas tank. A low fuel supply and no baggage may be desirable from a weight standpoint, and a minimum of fuel may reduce fire hazard following an accident, but a low fuel level also reduces the fuel head to a minimum. Lowering the tail, as in a climb, reduces its further on most aircraft, and the surge of fuel to the rear of the tank under the acceleration of the takeoff and pull-up may reduce the flow from a marginal tank installation to zero. Take whatever steps are necessary to assure that the fuel flow will be at its MAXIMUM during that most-critical first takeoff. Fuel feed from various attitudes and low fuel levels can be checked out on subsequent flights by burning off fuel in the air after the takeoff has been made with a safe load.

Caution should be the keynote of the full-flight program. Even though the plane may have plenty of zip, do not pull up into a maximum-rate climb. Climb out at a shallow angle at something less than full power, especially if the engine is new or just majored. While the plane should be flown outside the traffic pattern to check it out for stalls, etc., it is a good idea to first circle the airport several times in case the engine overheats or troubles show up that would prompt an immediate landing.

First Landing

Before making the first landing, take the ship to a safe altitude outside the airport traffic pattern and practice glides and stalls. By noting the airspeed at the stall, whether the airspeed indicator is accurate or not, the pilot will have a reference number to guide him during the critical first landing. Also, by flying under power and in a glide, he will establish the change of trim that results during the change from power-on to power-off flight. Just how much can be accomplished on the first out-of-the-pattern flight will depend on how the flight goes, the pilot's experience, and other variables. In general, this first flight checks out the powerplant and the stability and control of the plane. Testing for actual measurement of performance, high speed and flutter should be undertaken in gradual stages.

Do not be afraid to make several passes at the field before trying the first landing. Rate-of-sink figures obtained at altitude from the Rate-of-Climb instrument or altimeter and watch do not mean much to a pilot until visually translated into the rate at which the ground comes up to meet the plane on the landing approach. This rate will be quite a surprise to one making his first flight in a single-seater homebuilt after flying only Cubs. The rate of sink on the approach can be reduced by **flying the** plane in nose high under power at a speed safely above the stall speed determined by the stall tests. On power-off landings, carry a bit of excessive speed. If the plane, especially a small one, is brought in too slowly it will not "flare" close to the ground when the nose is raised, and the increased drag will only increase the rate of sink and result in a high bounce.

Minor Adjustments

In almost all cases, a need for minor adjustments will become apparent during the first or early flights. Take care of these as soon as they appear, especially looseness in controls and fittings. An accumulation of loose items, any one of which would be insignificant by itself, can become catastrophic. Hold off on the high speed and spin tests until the plane is well proven at lower speeds and the initial stretching of wires and cables has been taken care of.

While it is seldom a problem on low-speed planes (under 150 mph), each new design should be given a careful check for control surface flutter. This is accomplished by displacing the various controls one at a time to their maximum displacement and then letting go of the stick or rudder pedals. The surface should return promptly to the neutral position and stay there. Susceptibility to flutter on unbalanced surfaces increases with speed, so if the characteristic is present, it will not catch the pilot by surprise if he starts flutter testing at low speeds and gradually works toward maximum. By detecting it at the lower limit, the pilot can easily kill off a flutter as soon as it begins merely by reducing speed.

While true control surface flutter is an aerodynamic and weight relationship that is activated at a certain critical airspeed, an entirely different flutter can be introduced at low airspeeds by engine vibration or other factors if hinges are sloppy or control cables are under-tensioned. Spin testing should be done AFTER the flutter testing, since recovery from the spin is made by diving and speed can build up very quickly on a clean ship in even a very short dive.

Calibrating the Airspeed Indicator

Calibrating the airspeed indicating system can sometimes be a frustrating job, especially on a new design. The accuracy of the system depends on such variables as location of the pressure pickup, its size, angle relative to the airstream, tightness of the system, location of the static port, and other unknowns. Accurate calibration for the full-speed range is difficult to obtain on any airplane because of the nose-high (high angle of attack) attitude of the ship at low speed, and the nose-down attitude at high speed.

For the very low end of the scale, the plane can be clocked by a car running down the runway under no-wind conditions (get permission from the airport manager before using this method). At the cruising and high speed end, flying formation on another plane is practically useless unless the system in the other plane is known to be accurate. Besides, unless plane-to-plane radio is involved, communication between the planes can be difficult.

The easiest way is to make straight-line timed runs over a measured course. Make two passes, one each way over the same two points, to cancel out wind errors. A strong cross wind will result in erroneous data because of the need to fly in a "crabbed" attitude, thereby increasing the actual distance flown over the point-to-point distance. The course should be flown at as steady a speed as possible. Set up an engine r.p.m. value for each run and write it down. Opposite that write the average airspeed reading for that setting. DO NOT CHANGE THE THROTTLE SETTING! Hold the airspeed and the r.p.m. constant with the control stick. Allowing the nose to rise will decrease both r.p.m. and airspeed, so it is obvious that there should be no altitude change during speed calibration runs. Opposite the indicated airspeed obtained, write the elapsed time for the upwind and downwind runs over the particular distance. The actual speeds for a number of runs can be calculated later on the ground. Having the throttle setting for each run recorded allows repeat runs to be made at the same power increments to check the original readings against those obtained after alterations have been made to the airspeed system.

It should be possible to adjust the rig and trim of those homebuilts that have docile flight characteristics approaching those of production lightplanes to the point where the plane will fly straight and level at cruising speed without the pilot touching the controls. However, a plane that won't do this does not necessarily have anything wrong with it. Some ships, like certain aerobatic types, racers, and little bombs with high wing loading just don't have this capability. They are a handful to fly, and the pilots know it before they even climb into them.

8. Legal Problems and Paperwork

Pilots who build their own airplanes are a special breed — they are born and not made. Of course, they are "made" to the extent that someone has to teach them to fly, but the personal characteristics of patience, perseverance, and attention to detail are inherent, not acquired. These fundamental characteristics also account for one of the most amazing facets of the homebuilt movement — the almost complete absence of "hot rodding," and the extreme rarity of violations filed against owner-builders for deliberate infractions of the civil air regulations. Apparently the same characteristics that give a person what it takes to labor for a year or more transforming sticks and metal into an airplane also endow him with conservatism, a healthy respect for law and order, and an appreciation of his own limitations.

Under civil air regulations, the legal status of the homebuilt airplane, officially defined as "Amateur-built," is considerably different from standard commercial aircraft. And some of the licensing and paperwork procedures are entirely unique to the homebuilt. The FAA recognizes four different classifications of aircraft: STANDARD, LIMITED, RESTRICTED, and — the one with which we are concerned — EXPERIMENTAL.

There are various sub-classifications under Experimental: *Research and Development,* which is for bona-fide factory prototypes during their testing prior to certification as Standard; *Demonstration,* which is pretty much the same thing, or else to demonstrate special uses of an aircraft in a category for which it is not certificated; *Racing,* which is self-explanatory; *Exhibition,* which covers special aerobatic show types and non-standard restored antiques that are flown mainly at air shows; and *Other,* which catches anything that doesn't fit into any of the other spots including *Amateur Built,* which is actually second on the list but left to the last here for purposes of discussion.

What Is "Amateur-Built"

Under the present interpretation of the definition, an Amateur-built airplane is just that — something that the builder actually constructed himself largely from raw materials. When the classification was new, quite a number of alleged "homebuilts" were created by assembling cut-down components of a number of standard production designs. Those certificated as amateur-built before the FAA crackdown can continue to operate, but no new ones can be made that are basically assemblies of production airplane parts.

Smaller components, such as landing gear, engine mounts, fuel tanks, trim mechanisms, etc., can be used on a true homebuilt, however. If a builder takes a standard type like a Piper Cub and slips the wings or puts in a larger engine than the certification allows, the plane will be jerked out of the "Standard" category and put in "Experimental," but not Amateur-built. There has to be justification for the issuance of an experimental certificate. The FAA established the Amateur-built classification for "Recreational and Educational Purposes" only, and hopping-up a standard type does not come under this definition, recreational though it may be.

The clipped-wing Cub would hardly be a racer, so "Exhibition," if the plane were to be used for aerobatics at air shows, would be the only other logical classification. However, the catch is that the plane can be used ONLY for the defined purpose and not for casual joy-riding. Many builders, or "Converters," are not aware of these limitations when they start modifying a standard type, serene in the belief that they will be able to use it as a plaything. Recreational flying is both the purpose and the justification for the true amateur-built airplane.

The words "Amateur-built" are unfortunate in one way in that they imply a non-professional background and sub-standard design and workmanship. While this is often true, many of the so-called "Amateur" designs are worked up by top engineers in the airframe industry, and the craftsmanship is frequently of a far more exacting quality than a factory could afford to use on a production-line lightplane. While the designs and the workmanship may sometimes be very "Pro," the use of the plane itself is very definitely amateur, both by definition and regulation.

Homebuilts CANNOT be used for revenue purposes; i.e., you cannot rent one out or use it in connection with a business, although you *can* use it for personal transportation when travelling on business, or for non-pay appearances at air shows. Normal use of homebuilts is for DAYLIGHT VFR (Visual Flight Rules) flight only, although some fully-equipped models are approved for night operation.

Can Students Fly Them?

To answer an oft-asked question, student pilots CAN fly solo in home-builts provided their Student Permits (FAA Form ACA-340) are signed off for that particular model by a qualified flight instructor. Note that the word is MODEL, not homebuilts in general. Student permits have to be endorsed by an instructor for each different airplane model that the student checks out in. Checking out in a single-seater, of course, is largely a matter of a cockpit check after the instructor is convinced, on the basis of previous experience with the student in a two-seater, that he can handle the single-seater.

Special problems have existed for student-pilot builders of rotorplanes, principally the Bensen Gyrocopters, for the signature of an instructor with a rotorplane rating was required. Such instructors are quite rare in many regions, and even where they are plentiful, 'copter time at about $100 an hour isn't going to encourage Gyrocopter owners to buy a check ride. In recognition of this fact, the FAA ruled, and announced at the 1962 FAA convention, that only an airplane-rated instructor's signature was needed to authorize student pilot solo in an amateur-built rotorplane.

The paperwork involved in getting both airplane and pilot cleared to fly looks really formidable when the total is considered, but it is not nearly as bad as it seems. It is a relatively minor hurdle compared to the over-all job of building a plane or learning to fly. Most of the griping about paperwork in aviation, at least in the amateur field, comes from those who think they have everything taken care of and only then find out that there are still several forms to go through that they knew nothing about. The following paragraphs deal briefly with every bit of paperwork that the amateur builder-flyer will be concerned with between the start of work and the end of his first year of flight on the bird.

If you are new to all of this — especially when it comes to dealing with the FAA — it is a very good idea to call them up or visit them and talk the various steps over rather than to start filling out papers and sending them in cold just a few days before you need the results. Some of these things take time, and too little or inaccurate information can cause long delays.

Make And Model Designation

Every plane has one of each, and the amateur builder will have to dream up his own. It is customary to name an original design after the designer-builder, as Jones Model 1. "Sport" and "Special" are so widely used by homebuilders that they are practically aeronautical cliches. Frequently, a new owner remodels an existing homebuilt and gives it a new name. If the plane is built from purchased plans without alteration, it should be called by the proper name, as the well-known Smith "Miniplane." If the design is altered, it is a good idea to combine names, as Jones-Smith "Miniplane" for one that Mr. Jones modified considerably.

Serial Number

Again, every plane has one. There is no established system for this, and FAA is happy as long as the number doesn't duplicate that of a similar airplane. Most builders call their first plane, especially if of their own design, Serial No. 1. For purchased-plan designs this becomes a problem, except in cases where the seller assigns a serial number with each set of plans he ships out.

Nameplate

Something else the plane must have. This should be of metal, with the data metal-stamped on. Minimum information includes *Make* and *Model, Serial Number, Name* and *Address of Manufacturer,* and *Date of Manufacture.* EAA sells standard nameplates for homebuilts.

Logbook

This is not a government form, but every airplane must have two of them (Gliders only one) — one for the airframe and one for the engine. They must be hard-cover documents, not dime-store notebooks, and are for sale at most airports and at aircraft supply stores. These are laid out so that the required entries are self-evident. Sufficient to say that all flight time, mishaps, repairs, and inspections must be duly logged and available for inspection by the FAA. Since the entries are sworn statements required by law, falsifying logbook data is a misdemeanor.

While the information is not carried in the logbook of standard aircraft, since it is readily available elsewhere, it is a very good idea on a homebuilt to start the log with pertinent construction data such as the date started and finished, materials used, FAA inspections and sign-offs during building, and especially any information that a later owner might need: type of dope used, for instance, sources of non-standard materials, and even the type of oil used in the engine.

Initial FAA Inspection

Your local FAA Engineering inspector is required to check over your homebuilt before it is covered. If he approves what he sees, he will "sign it off" as OK to cover. The logbook is a good place for this notation, as it will be a permanent record. If certain items of the structure are of "Box" construction, the inspector must OK them before they are closed up, even if the rest of the structure is hardly started.

If the inspector does not like what he sees, he will suggest the necessary changes or improvements, and may want to see it again before giving his approval. This sometimes gets to be a sore point with many builders, but don't fight it. Airplane details are his job, and it is his duty to protect the public, including you, from mishaps that can occur from faulty design or workmanship. The rose-colored glasses of enthusiasm blind many an otherwise conscientious builder to small discrepancies that are immediately apparent to anyone not so personally involved in the job.

Application For Registration, FAA Form 500 B

This is obtained from the nearest FAA office, or a visiting inspector, and is filled out in duplicate at any time prior to the first flight of the plane, using the appropriate name, model, and serial number for the aircraft and the registration number assigned at no charge by the local FAA office. The original of the form is mailed to the FAA Examination and Records Division, at 621 North Robinson, Oklahoma City, Oklahoma, along with a $4.00 registration fee and both filled-out copies of Form 500 (see below). The yellow carbon copy is retained with the airplane until the validated Form 500 is returned.

If a special low registration number is needed because the plane is too small to take a long one, or something special is wanted for personal reasons, it can be had from the Registration Division for a $10.00 handling fee if the desired number is not already in use. For short numbers, two digits and a following letter, the request must be accompanied by an affidavit from your local inspector stating that the plane is too small or has special features that prevent it from using a larger number.

Certificate of Registration, FAA Form 500

Both copies are sent to Oklahoma City, and the validated original is kept in the plane after it is returned. If the plane is sold, the buyer starts the process again with another Form 500B, using the registration letter already assigned to the plane unless he want one of his own. Unless he renames the plane, the name and serial number will remain the same.

Bill of Sale, FAA Form 500C

This is filled out by the SELLER of the airplane or glider, and must be notarized on BOTH copies. The buyer sends the original to Oklahoma City and keeps the carbon.

Affadavit of Original Construction

In the absence of the Bill of Sale, Form 500C, a person who builds his airplane himself must establish his title to it in FAA records by submitting a notarized letter to the effect that he built it himself from raw materials and various used components. This affidavit (keep a carbon) is forwarded to Oklahoma City along with the Forms 500 and 500B.

Identification Drawings and Photographs

Since yours may be a one-only design, the FAA needs something by which to identify it for the records. A complete three-view drawing is preferred, but several photographs of the completed ship, taken from enough angles to fully establish the entire configuration, are acceptable when accompanied by data as to the dimensions and powerplant.

Final FAA Inspection

This is again performed by the FAA Engineering inspector. It is at this time that you will wish you had the first inspection actually signed off in the logbook if a different inspector is now on the job. Since the plane is now supposedly ready to fly, the inspector will probably be extra thorough, and will be looking for all the bolts you forgot to safety, etc.

Before he OK's the machine for flight, he may ask you to run the engine for him. He will definitely want assurance that you have complied with the requirement for running the engine for at least an hour on the ground prior to first flight at speeds up to full throttle. This test should be entered in the airplane logbook. The inspector may make his final inspection in your shop with the plane set up, or he may prefer to make it at the airport prior to the first flight, which he is supposed to witness.

Application for Test Area

Since homebuilts are restricted to a limited area not to exceed a radius of 25 miles from the base airport for their first 50 hours of flight (75 hours for those powered with non-certificated powerplants), application must be made by letter to the nearest FAA Air Traffic Supervisor for approval of a suitable area. Since this area must be off airways and away from congested areas, the full 25-mile radius will not always be granted. Sometimes the test area is quite distant from the base airport, in which cases a designated corridor to it is established. Assignment of the test area usually carries a time limit of anything from a few weeks to a year, and is, of course, for daylight and VFR conditions only.

Application for Certificate of Airworthiness, FAA Form ACA-305

This is the blue form that is filled out in single copy for the first airworthiness certificate for the plane and for all subsequent annual or other renewals of the airworthiness certificate. For homebuilts, check Item 2d, *Experimental,* and sub-category *Amateur-built* in addition to supplying the other required data as to owner, make, model, etc. With test areas approved, the form ACA-305 can be filled out as late as the day of the first flight if the inspector is willing to do the final paperwork at the test site.

Airworthiness Certificate, FAA Form 1362B

This is filled out by the *Engineering inspector* the first time a new homebuilt is ready to go, and by the *Safety Inspector* for subsequent renewals. Like the Form 500, the Form 1362B, must be in the plane at all times when it is flown. Actually, they should be there all the time, but some prudent pilots take them home, along with logbooks and other loose items, if the plane is kept at an unguarded field. Souvenir collectors liberate the darndest things sometimes. And missing papers on an airplane can spell big trouble for the owner.

Ordinarily, the Form 1362B is issued for a maximum period of one year for typical homebuilts. If the plane has unusual features that need proving, the inspector may decree a shorter time period. Some exhibition types are actually licensed on a day-to-day basis.

Operating Limitations

These are made up by the inspector, and pretty much tell you what you can and cannot do with your airplane. For the first 50 hours, these may seem unnecessarily stringent, but remember, this is the "Service Test" period in which the airplane must be proven, and cautious operation is in order. The initial restrictions can be modified at the successful completion of the test period. For two-seaters, one of the original limitations will be a prohibition on carrying passengers. In some cases, a second person can be justified as "essential crew," but the inspector has to be convinced. If you want to test at full gross weight, put a sandbag in the second seat.

Application for Modification of Limitations

This is another letter written by the builder, this time to the FAA Engineering inspector at the end of the 50 (or 75) hour service test period, which may have been completed by a relay of pilots in a week or stretched out by one man for a year or more. It simply states that the airplane has compiled with the time requirement and is ready for further demonstration to the inspector for his approval and modification of the initial restrictions, particularly the area limitation and passenger prohibition.

Certain limitations applicable to ALL amateur-built planes will remain in effect, however, such as "no flight over densely-populated areas, stay within continental limits of the United States, VFR flight only, etc." At present, homebuilts cannot cross the border into Canada except with special permission, but the problem is being worked on by both countries.

Passenger Warning

If the plane is a two-seater or better, the initial passenger prohibition can be removed at the end of the service test, but a placard must be installed as follows:

PASSENGER WARNING
THIS AIRPLANE IS AMATEUR-BUILT AND DOES NOT COMPLY WITH THE AIRWORTHINESS REQUIREMENTS OF STANDARD AIRCRAFT

EAA sells these, too.

Radio Station License, FCC Form 404-2

If your plane carries radio, and many homebuilts are using battery sets now that all airports served by FAA-operated control towers have required them since December 26, 1961, it must have a station license. The application, Form 404-1, is part of the same piece of paper—*obtainable from the nearest Federal Communications Commission office, not the FAA*. Like the FAA 500, the validated FCC Form 404-2 is returned and is carried in the plane.

Periodic Inspections

For homebuilts with an airworthiness certificate that is good for a year, a complete inspection of the plane must be conducted before that period lapses, and application made for a new certificate. It is illegal to fly a plane with a lapsed certificate. The inspection is written into the logbook by the builder or owner. On a standard-category plane and others, including most experimentals, this must be done by a certificated mechanic. But for the Amateur-builts, the owner-builder can do it himself. However, if the plane is powered by a certificated engine, it is a good idea to have a licensed powerplant mechanic go over it and sign it off.

When satisfied that the plane is still (or again) airworthy, call in the Safety Inspector, who will go over it himself. This time, and for subsequent re-certifications, the responsibility lies with the local FAA *safety* office instead of with *engineering* as on the initial certification and removal of service test restrictions. Re-certification of homebuilts must be done by the FAA, NOT by designee inspectors, as can be done with standard-category types.

If the inspector is satisfied, he will take the new Form ACA-305 which you just filled out for the occasion, pick up your old Form 1362B, and issue you a new one. He will then make his own entry in the logbook to the effect that he has found the plane airworthy and has issued a new certificate due to expire on a certain date. Standard-category planes have so-called "Permanent" airworthiness certificates that are re-validated after a designee or FAA inspector signs the re-validation in the log book. The log, then, becames the only paperwork indicating the current airworthiness status of the plane and MUST stay with the ship.

This procedure does not apply to homebuilts, but since many would-be builders are familiar with it they go under the mistaken impression that it does.

Affidavit of Non-modification

When applying for renewal of the airworthiness certificate on a homebuilt, the FAA requires an affidavit, actually a simple letter from the applicant to the inspector, stating that no major modifications or alterations have been made to that particular airplane since the last application. This is important, for most airworthiness certificates for homebuilts are now issued with a statement on the back stating that any modification of the airplane will invalidate the certificate.

Accident Report

While not pleasant to think about in connection with recreational flying, accidents do happen. Many pilots get themselves in trouble, not as a result of any violation of regulations that may have led to the accident, but in failing to file a proper report about it afterward or by hauling the wreck away before the FAA gets a look at it. Wreckage should not be moved until authorized by FAA, especially if anyone has been injured.

For standard-category aircraft, an accident report does not have to be filed if the damage is under $100.00. For homebuilts, however, ALL mishaps, however minor, must at least be recorded in the logbook and the FAA advised. Many builders neglect to do this, rationalizing that they don't want to bother the FAA or let them know that they are having troubles with their homebuilts.

These people do not realize that they are doing all amateur builders a disservice. The operating limitations on homebuilts exist partly because service life, maintenance, and accident data are not very extensive, so the limitations are applied to control a relatively new and unknown aspect of general aviation. As more information on the actual problem areas turns up, proper regulations can be formulated to correct them. Before this can be done, however, the necessary data are needed by FAA and by EAA as well for dissemination to other builders.

Personal Paperwork

Little need be said about the paperwork applicable to the pilot of the amateur-built airplane, since learning to fly is not a do-it-yourself project and there are established professional personnel involved to point the way through the paper snowstorm. The pilot needs at least a student permit properly endorsed by a qualified instructor, or a private or higher license with rating suitable to the type of aircraft.

However, a private pilot with only a single-engine landplane rating can SOLO a single-engine seaplane, be it standard or amateur-built, without a checkout in it, but he must have a rating appropriate to the aircraft, and be CURRENT IN THAT MODEL, meaning five full-stop landings in the past 90 days or a similarly-timed signed-off check flight by an instructor, before he can carry passengers in it.

In a homebuilt, he cannot carry passengers for hire even when he has a commerical license that enables him to do so in a standard type. If the homebuilt has radio, he must have a Restricted Radio Telephone Permit, FCC FORM 753-3, which is obtainable from the FCC practically for the asking. In flying on a student permit, the pilot must have his permit signed off for cross-country before taking the homebuilt more than 25 miles from the home field, assuming that the area limitations have been removed. A student pilot CANNOT make the first flight of a new homebuilt.

Miscellaneous

In the line of non-official paperwork, it is a good idea to keep a record of all purchases of material for your airplane, whether new or used. This will keep you informed on actual costs, will enable you to answer the inevitable question that you will hear dozens of times: "How much did it cost ya, Mister?," and will enable you to help friends who are considering homebuilts of their own or yourself when planning another.

Most important of all, however, it will enable you to establish a true cash value for your machine when the tax assessor comes around. If he is not experienced in evaluating aircraft (some states tax airplanes as personal property while others use an excise tax) he may arbitrarily assign an unfairly high value in the absence of substantiating figures. Also, sales slips show that state sales taxes have been paid on the raw or used materials, another concern of the tax people.

Try, too, to keep at least a rough log of your working time. The next most-asked question from the spectators is: "How long did it take you to build it?" It's nice to be able to tell them, but few builders can come up with specific figures in terms of actual man-hours. Most of the answers are generalities expressed in years, or in some cases, months.

9. Costs

Answering the question "How much does a Homebuilt cost?" is practically impossible without a lot of hedging and qualification. The costs simply cannot be calculated on the same basis as that used by industry, with amounts figured in for labor costs, overhead, taxes and other considerations. Home building of aircraft is recreational, and many of the items that must be added to the price of the commercial product are either not considered at all or are charged off to other things. For those who have to rent a shop in which to build their planes, this rent, and the heat and light bills, are very real parts of the overall cost. Costs go up, too, when specialized jobs like welding and machining must be farmed out, or special tools rented.

However, varied as these costs are, they cannot be added to the value of the final airplane when sitting alongside another built by someone who did all his work in his own garage with his own tools. The comparative worth of two homebuilts of the same model will be determined by the more significant factors of craftsmanship and equipment, and even the color scheme can enter into it when a tax assessor who knows nothing about airplanes has to put a valuation on one.

In general, the cost of a homebuilt is considered to be the cost of the raw materials and purchased equipment at current market prices, plus such hired work as the cost of having a cheap run-out engine overhauled. Some homebuilts now flying are quoted as having unrealistically low initial costs because they were assembled from cut-down components of a junked production airplane such as a Piper Cub or an Aeronca. Since FAA has now prohibited this practice, such machines should be re-evaluated in terms of what it would cost to build them up from raw materials.

For one who has to go into the open market and buy all of his raw materials and used equipment, $1,000 is about the minimum price that one can expect to meet when building a simple single-seater powered with used 65-h.p. certificated airplane engine. The author's all-wood "Fly Baby," described in Chapter 11, cost $1,050; $375 for all the new raw materials such as plywood, aircraft grade wing spars, sheet metal, nails, glue, fabric, and dope, and $675 for such used hardware items as a newly-majored 65-h.p. engine, metal propeller, wheels, brakes, fuel tank, and

Lowest-cost form of homebuilt, which, unfortunately, is no longer legal. This Nelson N-4, believe it or not, was once a Piper "Cub." Builder Ray Nelson cut down the major components of the factory-built ship and rearranged them into this racy sportplane. (Photo by Peter M. Bowers, 1956)

instruments at 1960 prices. "Extras" such as electrical systems, radio, and additional instruments can add considerably to the initial cost. Unless one goes hog-wild or buys everything new, the top cost should not be much over $2,000.

Raw Material

Raw material cost will vary greatly with the nature of the materials, wood being by far the cheapest when using marine and exterior plywood. True American-made aircraft grade plywood, well over a dollar a square foot, is practically out of the picture for use in any quantity. A complete set of aircraft grade spruce spars for a small biplane like the Mong Sport costs $39.50 at firms that make them up. The spars for the bigger EAA biplane are $75.00 at the same places, with the Cougar/Tailwind monoplane in the middle at $49.25.

Steel tubing in the sizes used in homebuilts runs from $.40 to around $1.00 per foot; while sheet aluminum, when it can be found in surplus yards, may run as low as $.25 per pound and on up to $.50 per pound or a bit more as new material. The cost of fabric covering and dope varies greatly, from cheap unbleached muslin—which is still used to some extent at under $1.00 per yard—to the new Dacrons exemplified by the special commercial Ceconite at nearly $3.00 per yard. Long-term cost is hard to evaluate here, as the Ceconite will outlast the muslin by about four to one, eliminating the need to recover. Dope ranges from the lowest-cost clear nitrate at $2.50-$3.00 per gallon to pigmented Butyrates at $10.00. The finish of the plane can add greatly to the final cost when one builder prefers a 15-coat hand-rubbed finish on Ceconite to someone else's bare-minimum three or four coats on Sears-Roebuck or J. C. Penney Dacron Taffeta.

Engine Costs

Except for Volkswagen conversions and the Lycoming ground units, practically all engines used in the average homebuilts are second-hand stock aircraft types. Price varies in fairly direct proportion to power and condition, run-out 65's in need of major overhaul going for $150 or less while newly-majored engines of the same power bring $350 to $500—and good low-time 85's and 90's reach highs of $600.

While these prices may sound high for used equipment, FAA regulations guarantee that they are in virtually new condition and with normal treatment will not need overhaul again for at least 1000 hours, which is five or more years of normal flying for most homebuilts.

A major difference between sport planes and sport cars is that the planes do not compete with each other on the basis of performance-in-class, where tinkering with the engine can make all the difference. The modern aircooled airplane engine in the 65-150 h.p. range used by the homebuilts is a remarkably reliable and trouble-free mechanism. And the builder is content to leave it strictly alone for as long as possible, the only upkeep cost being oil changes and an occasional replacement of spark plugs.

Operating Costs

Cost of ownership and operation is something entirely different, and is largely an individual matter. Some owners rent hangars at going commercial rates from $20 to $50 per month, others tie the ships out on the line at $5.00 or $10.00, and a very few fold the wings and haul their birds home to their own garages. Regardless of the sometimes painful cost, keeping the plane indoors is a saving in the long run. The covering, finish, and sometimes the equipment suffer from exposure when tied out for long periods, and a recover may be necessary in one-third to one-half the time that could be expected for a machine kept indoors.

Actual operating costs in terms of gas and oil are very low. The small 65-h.p. engines burn around four gallons of gas per hour, and even the 150's use between seven and ten depending on power setting and altitude. Oil consumption is about equivalent to automobiles of the same power. Unfortunately, airplanes have gotten an unfair reputation for expensive operation because of the commercial operators' practice of charging for rental by the hour, which has become the standard for calculating all aircraft operating costs.

While this works well for purely commercial operations, where all the costs of the entire operation—salaries, overhead, taxes, initial cost, depreciation, and profit—must be charged against the working time of the revenue-producing plane, it cannot be used to determine the true costs of strictly recreational equipment such as sportplanes or pleasure boats when they are owned or have been built by their users.

Many of the tangible commercial considerations disappear and are replaced by such intangibles as the mere satisfaction of creation and ownership, an object for enjoyable tinkering and improvement without a bit of operational time put on it, and so on. If commercial rental airplane costs were re-stated in terms of cents-per-mile as are rental automobiles, the comparison would be much more favorable to the airplane than it is now. And if powerboats were evaluated in cents-per-mile, or even the automobile's miles-per-gallon, the airplane would look still better by comparison.

"It costs too much" is seldom the critical factor in keeping a person from building his own airplane. The project must be evaluated like any other technical hobby such as Hi-Fi, photography, or a boat or car. The return on the investment is largely a matter of the individual. The relatively high turnover in completed homebuilts is not because of excessive cost, either. Frequently the owner outgrows relatively simple equipment and wants something with better performance, or, conversely, he finds that the machine he owns is a bit beyond his capabilities.

Often a change of jobs and subsequent relocation forces a sale. If you buy a homebuilt that has been advertised for sale, you may get any of a great variety of interesting answers to your question of why it's on the market, including "She said it was it or me," but it will hardly ever be "I can't afford it."

10. Ground Transportation

Very little attention is paid to a particular problem area that has special application to homebuilts—ground transportation. Where most factory-built types virtually fly out of the back door, the homebuilder has to get his product from home to a sometimes very distant airport. In many cases, especially with antiques and replicas, trips to even moderately distant Fly-ins are made by trailer. For these latter types, the problem is recognized in advance and provision usually made for a permanent and suitably equipped trailer. The builders have taken their cues from the soaring fraternity, in which every glider has its own trailer that enables it to be loaded and secured very quickly—yet still be capable of standing the rigors of a long trip and heavy weather.

The average homebuilt ready for its first trip to the airport seldom gets such deluxe treatment. More often than not it is made to serve as its own trailer, with its tail lashed to the rear bumper of a car and the wings tied alongside the fuselage or on top of the car. But while this procedure may be all right for the airplane, if the trip is short and the weather mild, it may be frowned upon by the highway patrol. In some states they insist on a one-way permit for such operations.

Because of the EAA design contest (see Page 118) and the emphasis on folding-wing designs that can be stored at home, the surface transportation problem has received some long overdue attention and the haul-home practice can be expected to increase. The initial contest requirement for transportability at 40 mph resulted in some designs that could be towed tail-first on their own wheels and be licensed as trailers. A trailer hitch was bolted to the tail, and jumper wires hooked built-in brakes, clearance, and turn signal lights to the car battery. When the towing requirement was upped to 60 mph, however, the contestants and others almost unanimously went to transportation on trailers. These were either adapted from normal boat or glider types, or built specially for the particular plane.

Even at only 40 mph, there are many disadvantages to hauling planes appreciable distances on their own wheels. In the old days of plain bearings it was necessary to stop frequently and grease the axles, but modern sealed roller bearings are not a problem here. Brakes, however, are. Some expander-tube and disc types have a little drag, and in several road miles can become almost

The traditional way of getting most small airplanes from home to the airport. This Sorrell "Special" is tied to the tailgate of the station wagon and the small wing panels are stowed inside. (Photo by Peter M. Bowers, 1957)

A variation on towing the airplane—Bowers "Fly Baby" with wings folded and a special trailer hitch provided for fastening to car. Highway Patrol usually frowns on such procedure, but proper lights and wiring enable this combination to be licensed as a bona-fide trailer. (Photo by Peter M. Bowers, 1960)

red hot. Trouble can result if the wheels are just a little bit out of track, and since airplane wheels are usually quite heavily cambered to accomodate shock-absorber action, towing can wear them unevenly when the normal full weight of the plane is not on the wheels.

Single-leg spring-steel landing gear can be troublesome for towing since it can easily be out of track, and its "softness" can let the plane lean heavily to one side in a crosswind, especially if the wings are folded or tied alongside. A final disadvantage of this procedure is the rough treatment that the instruments get. They are usually located right over the wheels, where all the bouncing takes place, and get a terrific amount of up-and-down acceleration since the pivot point of the bouncing is back at the tie point to the car.

If this procedure is used, be sure that the joint to the car is flexible enough to accomodate both bouncing and tight turns, but still stiff enough to keep the plane from overrunning the car when the brakes are applied or when heading down hill. Be especially sure that the horizontal tail surfaces, if left on, will not hit the rear corner of the car in a turn. For short-range tail-first towing at low speed, it should not be necessary to batten the hinged tail surfaces to prevent their being damaged by wind and gusts.

Hauling Wings

Transporting the fuselage is relatively easy compared to the problem of hauling wings without a trailer. Only the smallest ones can be stacked with safety in the back of a station wagon or box trailer. If carried outside the car, covered wings are extremely difficult to stow; there are few points to tie to, the fabric is highly susceptible to punctures, the resistance to the wind is great, and they cannot be lashed tightly to the curved top of a modern car without a car-top carrier or special supporting structure. Uncovered wings or fuselages are not as big a problem; there is plenty of exposed structure to tie to, and the absence of covering greatly reduces the wind resistance.

Trailering is an infinitely more satisfactory procedure, but again there are specialized problems, the first one being the location of a suitable trailer. If you decide to build one and make steady use of it, good, because the trailer can be tailored to the particular plane. One-shot adaptations of boat, glider, or rental trailers generally require some superstructure to hold the plane securely. Glider trailers are generally best because they have cradles to take wings, and there is a secure vertical structure in front that the wing butt fittings can be attached to.

A flat-decked trailer is best, and if fitted with two simple wheel ramps and a boat winch, one person can load a fuselage or an entire folding-wing plane by himself. Some of the glider guiders have modified their special glider trailers to flat-decks that can be used to haul a variety of craft with a minimum of adaptation. Other flat-deck trailers can sometimes be borrowed from airports where they are kept available to haul in wrecks or to move seaplanes from hangars to the water.

Rental trailers are usually box types that are quite unsuited to hauling complete airplanes. If more or less continuous hauling is anticipated, the owner-builder should build up his own trailer, or have one built. Cost for even a "bought" 24-footer should be under $300, and a homebuilt one can be kept well under $100.

One of the advantages of small airplanes is the relative ease with which they can be transported when the covering is off and there is enough exposed structure to tie to. Fuselage is for a Volkswagen-powered French Druine "Turbulent" under construction by Joe Koskie of Seattle. (Photo by Peter M. Bowers, 1959)

Leon Trefft's "Contestor" on special short trailer at 1962 EAA Design Contest. Boat winch hauls plane up folding ramps. Extreme overhang is illegal in some states and presents interesting problems for towing in 30-40 MPH western crosswinds. (Photo by Peter M. Bowers, 1962)

"Fly Baby" with wings in normal folded position, loaded tail-first on a converted glider trailer. Note locks on ailerons and battens on tail surfaces. In spite of secure fastenings fore and aft, ship took severe beating from strong crosswinds on 2200-mile trip from Seattle to Rockford for 1962 Design Contest. (Photo by Peter M. Bowers, 1962)

Nose Or Tail First?

The location of the wheels on the trailer will generally determine which way the fuselage faces. With wheels far aft, the plane will have to face aft to keep the major weight near the wheels. Tail-first hauling on a trailer has the same disadvantages as towing on the plane's wheels—bouncing of the instruments and a need to secure the assembled tail surfaces with battens. If the plane is a folding-wing type with its wings folded and the whole assembly just rolled up on the trailer, the center or side area will be fairly high. For that reason the tie-down ropes, cables, or chains should go from the trailer bed to points fairly high on the fuselage to resist the side loads imposed by crosswinds.

If the wings are the folding type they may need extra tie points to hold them folded and rigid under bumps and gusts of wind. If the wings are carried separated from the fuselage they should be located as low as possible, and preferably bolted in place instead of tied with rope. Rope is very poor as it is difficult to tie tight in short lengths or to pull tight around the ends of surfaces. A little slack will let a wing slip and chafe.

For ties from strong points on the fuselage to the trailer bed, cable or chain tightened by clothesline turnbuckles which are safetied is the best for long hauls. Wheels should be blocked to prevent the slightest fore-and-aft motion on the trailer. One thing to watch for when hauling a plane lashed to a trailer bed but still on its own wheels is loss of air in the tires. This can let the main tiedowns slack off some and cause trouble.

If at all possible, set up the trailer so that the plane is pointed forward. It takes the air loads much better this way, and it is hardly necessary to batten assembled tail surfaces. The instruments fare much better as they are quite close to the pivot point instead of the bouncing wheels, and thus take much less of a beating. When tying a plane down on any trailer, do not make the fastenings at both ends of a long structure like a wing or a fuselage rigid. Any flexing of the trailer frame which might be transmitted to an airplane structure not designed to take loads of this nature can cause severe internal damage that may go indetected.

115

"Fly Baby" on earlier trip to Rockford, on different glider trailer that permitted forward mounting. Wings removed and carried in glider wing cradles, airplane wheels removed and axles fitted into chocks. This arrangement stood long haul across windy western states much better than method illustrated on page 113. (Photo by Peter M. Bowers, 1960)

An entirely different approach to the problem of transportation between home and the airport—L. D. Bryan's "Bryan II" folds its wings and goes down the road under its own power, which is legal since it is also licensed as a motor vehicle. (Photo by Peter M. Bowers, 1960)

11. The Top Homebuilts

One of the most remarkable things about the home-built airplane movement is the terrific longevity of the more successful designs. There seems to be no such thing as technical obsolescence or a personal attitude of disdain for "last year's model." If anything, there is a fascination for the older and more fundamental designs, with their structural and aerodynamic simplicity.

Part of this is sentimental and romantic—a deliberate attempt to recapture the spirit of those bygone days when flying was really a glamorous and even heroic business instead of the normal part of life that it is today. The terrific boom in the restoration of antiques and the building of replicas attests to this. The large number of biplane designs on the market represents a middle ground that manages to retain the more obvious features of the old while incorporating the advantages of modern equipment and materials.

The main reason, however, is more practical. The old designs that survive have proven themselves to be dependable. This is a point of prime importance to the builder who wants assurance that the piece of recreational equipment that he builds will return many hours of safe and trouble-free operation for his thousand or more dollars and hundreds of hours of time. While a small percentage of the builders are essentially gadgeteers and tinkerers who want to get the ultimate in performance out of a given powerplant, the great majority are satisfied with flight in their own creation, and care not a bit whether someone else's machine is a few mph faster.

In fact, just as in commercial aviation—where some of the really obsolete "clunks" of the late 1920's can still do some jobs better than the hot modern types, and manage to survive in a competitive business because of it—some of the oldest home-built designs have characteristics that outweigh some of the alleged advantages of the later designs. No one in the homebuilt movement thinks it unusual for someone to start building a 1930 Pietenpol Air Camper in 1962. In fact, two of these, altered only to the extent of using current airplane engines instead of the original Ford "Model A," were on hand at the 1962 EAA National Fly-in.

The Baby Ace

The Pietenpol is the most numerous of the really old designs, dating as it does to 1930. The little Heath Parasol, which pre-

The age of a design is no drawback to its popularity, and may even enhance it because of a reputation for reliability and performance earned over many years. This latter-day Pietenpol differs from the 1930 original only in later engine and modified wingtips and landing gear. See page 7 (bottom). (Photo by Peter M. Bowers, 1960)

ceded it, was much too marginal a flying machine to be acceptable in its original form today, although plans are available from EAA for the later type-certificated version. The Heath, however, was the inspiration for the Corben "Baby Ace" of the early 1930's, which saw limited production. In 1954 this design was extensively redesigned and modernized by Paul Poberezny and plans were published in *Mechanix Illustrated* magazine.

Plans, materials, and kits were also put on the market by a reorganized Corben company, and the Baby Ace became the first design available to postwar amateurs who preferred to build from proven plans rather than work up their own designs. Because of its head start in the field, with plans available for the cost of the magazine, the Baby Ace is the most widely-seen of the "Buy-the-Plans" homebuilts. One of its most appealing features is the leeway possible for incorporating the builder's own modifications and variations in equipment; this allows a high degree of individuality to a design that still remains safe and predictable.

Contemporaries of the Baby Ace, and actually on the plans market ahead of it, are the Stits "Playboy" and Steve Wittman's "Tailwind." Although a slightly heavier and considerably hotter design than the Baby, the Playboy is still a structurally simple and conservative design, and permits the builder wide latitude in modification—with powerplants from 65 to 125 hp. A demand for two-seater homebuilts resulted in a side-by-side version of the basic Playboy, but the change was not too popular. Playboys are the third most numerous of the "standard" homebuilts. Other Stits designs are also on the market, including the two-place "Skycoupe" and "Flutterbug" models, but the Playboy is favorite.

Number Two: The Tailwind

Like the Ace, the Tailwind design goes back to prewar days and Steve Wittman's "Buttercup," a fast high-wing cabin model built to use much of his racing plane know-how in a passenger ship. The postwar Tailwind, a side-by-side two-seater, appeared in 1953. It was greeted in stunned silence at first—no one believed that a two-seater with such a short span, 20 feet, could fly at all, much less perform effectively with full load.

The old master's touch was there, however, and the Tailwind, along with its near-perfect copy, the Nesmith "Cougar," has taken over the No. 2 spot in homebuilt popularity. Neither is a beginner's ship, however, nor are they suitable for extensive individual alterations. The builders are content to leave their tinkering to changing the shape of the rudders and to working out an infinite variety of nose cowling shapes.

Biplanes

Biplanes re-entered the scene in the middle 1950's with the Meyers "Little Toot" and the slightly simpler Mong "Sport." Both are metal-fuselage single-seaters of extremely clean lines. These were followed shortly by the Smith "Miniplane," a much more gadgety and old-fashioned-looking design that caught on fast and already outnumbers the combined total of Meyers and Mongs. While the Miniplane is a tight fit for larger pilots, the EAA Biplane, inspired by the Gere Biplane of the early 1930's, is somewhat larger and roomier, and is more in the docile Baby Ace class of handling characteristics. Because of this, it is expected to become the most popular beginner's biplane, especially with plans selling at only $10.00.

While most of the early postwar homebuilts were built by people with enough aircraft experience to be able to work with steel tubing and sheet metal to aircraft standards, no designs were available that were expressly laid out with the rank beginner in mind. When it became obvious that the movement was going to grow even more rapidly, the EAA proposed a design contest in

Top photo shows original Corben "Baby Ace C" as marketed in 1955; middle photo is improved "D" version, while bottom is a highly original modification. Use of closed or open cowlings, wheel pants, and other refinements are largely an individual matter.
(Photos by Peter M. Bowers, 1960)

1957. Under this the membership was invited to design, build, and submit designs for a low-cost, easy-to-build, and easy-to-fly folding-wing airplane for judging in a competition to be held in connection with the National Fly-in in 1959. The purpose of the folding-wing feature was to permit the plane to be stored in a standard home garage instead of at the airport.

While many entries were received, it was obvious that few if any would be finished by the deadline, so the contest was postponed to 1960. Again, many entries were promised but only two appeared. Since neither had been able to accumulate the required 50 hours of flight-test time required for removal of the initial FAA restrictions, or even the 15 hours time to which the contest requirement had been reduced, the contest was postponed a further two years to 1962.

Within two months of contest time, 18 designs were promised. But again the final results were disappointing, in that only six appeared. Three, including a really fine open-cockpit two-seater which had not been designed for the contest but which generally came within the specifications, were eliminated for various reasons that rendered them unsuitable for the objectives of the competition. One, Leon Treft's "Contestor," (see page 113), was definitely on the right track. It was an all-wood design of extremely simple construction, but unfortunately went too far in the direction of simplicity and the use of stock sizes of commercial wood to the point where performance was compromised by the use of too-small wings.

The Winning Fly Baby

The winning design was the author's all-wood "Fly Baby," which was one of the two competitors to appear at the 1960 contest. Cost was kept down by using commercial marine and exterior plywood for much of the structure, including the wing ribs, and using Dacron Taffeta, obtainable in department stores, for covering. Every effort was made to keep the structure extremely simple, almost to the point of crudity, so that anyone getting good grades in a high school woodshop course could build it without difficulty.

Special skill requirements were reduced by keeping such operations as welding to an absolute minimum. There is so little welding, in fact, that the builder can cut out his materials and take them to a professional shop for a short, and therefore low cost,

Top view shows a standard Stits "Playboy," with moderate and extreme variations following. Most popular modification appears to be addition of wing area. Note flaps on 3131C. See Page 17 for another modified "Playboy" with special equipment. (Photos by Peter M. Bowers, 1960-1962)

Normal configuration for two-seat Wittman "Tailwind," W-8 side-by-side two seater built by Jim Narrin of Flint, Michigan. These and the nearly-identical "Cougars" are the most popular two seaters on the market today. (Photo by Peter M. Bowers, 1960)

Meyers "Little Toot" is an excellent aerobatic biplane designed to cater to different structural preferences, either sheet metal or welded steel tube fuselage. This particular one, built by Arlo Schroeder of Newton, Kansas, is probably the best known because of distinctive military color scheme copied from famous Curtiss P-6E "Hawk" fighters of 1932-35, hence the name "Hawk-Pshaw." (Photo by Peter M. Bowers, 1962)

job. A careful check on the building time indicates that the original Fly Baby was built from scratch in 720 working hours, which is certainly as short a time as anyone can expect when building a complete airplane.

Aerodynamically, all compromises were made to the advantage of easy flying characteristics. High-speed performance was sacrificed for improved performance at the other end of the scale,

Volmer Jensen's "Sportsman" flying-boat amphibian, the first seaplane or amphibian design available to the homebuilder. Wood hull of original design, wings and horizontal tail from commercial Aeronca Champion or Chief airplane. (Photo by Peter M. Bowers, 1963)

The first of many—the 90 HP Smith "Miniplane" prototype. While those who build from the plans have made few modifications to the airframe, powerplant installations have ranged from 65 to 125 H.P. (Photo by Peter M. Bowers, 1960)

giving it "over-the-fence" characteristics matched by few other homebuilts. These were obtained primarily by using long-span wings of generous area, the span being 28 feet, larger than most other homebuilts including the two-seaters.

How well these features pay off on Fly Baby is shown by the fact that it has sufficient reserve lift to permit it to tow gliders (prohibited since June 28, 1962, for all but standard category aircraft) or to be fitted with twin pontoons, operations virtually impossible with other 65-85 hp homebuilts. Ceiling with 65 hp is 15,000 feet, and rate of climb with 85 hp is over 1,000 feet per minute. Landing speed is approximately 46-48 mph and cruising speed is 95-110 mph, depending on the engine and propeller. An unusual feature is the adaptability of the fuselage to the use of an interchangeable set of biplane wings to further increase versatility.

To be perfectly honest about it, Fly Baby is another of the "Latter-Day-Antiques." The designer had several years of experience with the two Story Specials, built by Tom Story of Beaverton, Oregon, who had developed them from George Bogardus' "Little Gee Bee," which he had helped design, and which *itself* was a development of the prewar Les Long "Longster" known as "Wimpy."

All of these machines featured long-span wings with wire bracing to a rigid landing gear. Fly Baby followed this general pattern but used wood instead of steel tubing for cost and simplicity reasons. Also it went to the even older designs of the Heath "Baby Bullet" and Howard "Pete" racers for a still simpler landing gear, where the underwing brace wires ran directly to the ends of the axle instead of to a secondary structure behind the wheels. While the simple straight-axle configuration without shock absorbers did not cause much comment, the fact that the supporting struts were made of wood was a severe jolt to practically everyone. Even the all-wood Pietenpol had switched to steel tubing for this unit in the early 1930's.

Using wire bracing on the wings was not just copycat technique. Most of the modern low-wings that are not cantilever construction use struts between the top surface of the wing and the upper longerons, but for the most part they are short-span designs with the struts installed at a rather steep angle. Because of Fly Baby's long span, this angle would be quite flat; there would also be much aerodynamic interference in the wing-strut intersection area, where even some of the short-span types have had trouble.

Wires reduce drag in this critical area considerably, and at ten cents per foot are infinitely cheaper than heavy-wall streamlined tubing at $3.00 per foot and up. Some pilots who are unused to wire-braced designs are somewhat shaken by seeing the upper wires slack off a bit under high flight loads, and this characteristic

For those who can't fit into the "Miniplane," there is the EAA Biplane. This prototype, with EAA president Paul Poberezny in the roomy cockpit, was built by students at St. Rits's Technical High School, Chicago, as a class project. (Photo by Peter M. Bowers, 1962)

One of the raciest looking biplanes ever built, a modified "Knight Twister" designed by Vernon Payne and built by Tony Sablar. Small wing area keeps standard version out of the average sportsman's class, but more docile long-wing versions are available. (Photo by Peter M. Bowers, 1962)

Sporty-looking and original "Tuholer" by Tony Spezio, of Oklahoma City, entered in EAA Design Contest even though not developed for it, is a serious challenge to the virtual monopoly of the high-wing cabin designs in the two-seater field and can be expected to encourage other open designs. (Photo by Peter M. Bowers, 1962)

takes a bit of getting used to. Other features, such as the shape of the rudder and wingtips, and even the layout of the color scheme, are sentimental carry-overs from the author's "Rebel" gas model airplanes of 1937. Even the name "Fly Baby" is taken from his record-holding Class A/B gas model of 1940.

Second Place: The T-40

Second place in the contest went to Gene Turner's all-wood cantilever low-wing single seater, the T-40. While almost as simple as Fly Baby structurally, it was of considerably higher performance, with a few complex features that made it more suitable for a beginner's second project than his first.

Third place went to Leonard Eaves' folding wing "Cougar," the only two-seater and steel-tube fuselage design to place. While not of his own design, this adaptation qualified because permission to use it was granted by the originator. Eaves' contribution was the folding wing, which brought this established design completely within the contest requirements.

By contest rules, the plans for these designs are available from the originators and are listed in the plans catalogue of Chapter 4. Unfortunately, space considerations prevent including detailed descriptions of many other good designs that are on the market. Their deletion should not be taken as an indication that they are in any way inferior to those described. A feel for their size, weight, power, and performance can be obtained from the specification tables at the end of Chapter 5, and the sources and the cost of the plans can be found in the listings of Chapter 4.

* * * *

There you have the facts on the homebuilts. Buy those plans, or dust off the drawing board and sharpen the pencils. What's to keep you from building that airplane now?

Author's all-wood "Fly Baby," winner of 1962 EAA Design Contest, owes its docile flight characteristics primarily to generous 28-foot wing span. Note "Hands-Off" flight. Object under belly is not auxiliary fuel tank but a streamlined suitcase. (Photo by Dale Weir, 1962)

Gene Turner placed second in EAA contest with T-40, a snappy all-wood design considerably faster and slightly more complex structurally than "Fly Baby." Judges ruled it more suited to relatively experienced builders than to beginners. (Photo by Peter M. Bowers, 1962)

Third place in EAA contest won by standard Nesmith "Cougar" with folding wing feature developed by Leonard Eaves of Oklahoma City. Unlike most "folders," this one can be taxied under its own power when folded. (Photo by Peter M. Bowers, 1962)

AIRCRAFT MATERIAL SOURCES — MISCELLANEOUS

Ace Aircraft Mfg. and Supply, McFarland, Wis.
   Tubing, Plywood, Spruce, Hardware, etc.

Advance Aircraft, Inc., Educational Dept., 161 E. Grand Ave., Chicago 11, Ill.
   Correspondence Course Light Airplane Design, Drafting

Aerial Blight Control, Inc., Box 224, West Bend, Wis.
   Tubing, Wood, Parts, Engine Overhaul

Aero Marine Craft, Inc., 216 Coleman Ave., Long Branch, N.J.
   Custom Styled Control Wheels

Aero Publishers, Inc., 2162 Sunset Blvd., Los Angeles 26, Calif.
   Books

Aero Supply & Equipment Co., Box 96, Stanley, Kans.
   Plywood, Spars, Cap Strip, Glue

Aero-Trades, Inc., Harrisburg Airport, Harrisburg, Pa.
   Aircraft Service and Repair

Aircraft Miniatures, Delavan, Ill.
   Trophies

Aircraft Spruce, Box 521, Encino, Calif.
   Spruce Spars

American Surplus Trading Co., 332 Canal St., New York 13, N.Y.
   Crash Helmets

ASCO, P.O. Box 483, Hermosa Beach, Calif.
   Pitot Tube Covers

Aviation History Publications, Box 624, Concord, Calif.
   Books

Benjamin Sales Co., 3235 Fenkell, Detroit 38, Mich.
   Aircraft Turbines, Dzus Fasteners, Bolts, Nuts, Screws, Fittings, Pumps, Instruments, Switches

George C. Benson, 24850 E. 2nd St., San Bernardino, Calif.
   Plastic Laminating (paper)

B & F Aircraft & Supply, 6141 W. 95th St., Oak Lawn, Ill.
   Tubing, Plywood, Instruments, Propellers, Engines

M. J. Bowe, Box 3086, Lennox, Calif.
   Wheels, Mags

Troy Bowers, Jr., 310 Ave. S., Lubbock, Texas
   Fiberglas Wheel Pants

Bruce Billings, 9725 S. Main Rd., Rockford, Ill.
   Aircraft Paintings

Capitol Copter Corporation, 909 Johnson Pkwy., St. Paul, Minn.
   Rotor Blades

Celon Industries, Lawrence, Mich.
   Listings, Parts

Cleveland Model and Supply Co., 4506 Lorain Ave., Cleveland 2, Ohio
   Model Airplane and Railroad Kits

Cleveland-Peerless Antique Models, 4500M 146 Lorain, Cleveland 2, Ohio
   Catalog—Old-Time Plans

Columbia Airmotive, P.O. Box 654, Trautdale, Oreg.
   Wheels, Brakes, Hardware, Instruments, Parts

Cooper Engineering Co., Box 3428, Van Nuys, Calif.
Ceconite Fabric, Air Fibre Fiberglas

Cooper Industries, Inc., 2149 E. Pratt Blvd., Elk Grove Village, Ill.
Tubing, Wood, Fabric, Dope, Instruments, Parts, etc.

Copter Sales, 14535 NW 13 Ave., Miami 68, Fla.
Gyrocopter Float Planes

Design Sheet, P.O. Box 54, Eaton Rapids, Mich.
Lightplane Design Sheet

Engineering Research, R. R. 3, North, Kendallville, Ind.
Machine Work and Welding

Fibreglas Aircraft Accessories, Rt. 2, Box 187, Forest Grove, Oreg.
Fibreglas Wheel Pants

W. J. Fike, Box 83, Anchorage, Alaska
Design Booklet

Flight Fabrics, P.O. Box, Abington, Mass.
Dacronite H.S. Synthetic Fabric for Aircraft Covering

GN Aircraft, 355 Grand Blvd., Bedford, Ohio
Cap Strip Stock, Spars, cut-to-order wood kits.

Tom Gunderson, Twin Valley Airport, Twin Valley, Minn.
Welding

Harrod's Aviation Supplies and Sales, 2851 N.W. 159th St., Opa Locka, Fla.
Wheels, Brakes

Ray Hegy, Marfa, Tex.
Wheels, V-W propellers

E. F. Howington, 3360 N.W. 37th St., Miami, Fla.
Wheels, Brakes

Johnstone Aviation Co., R. R. 2, Elkhart, Ind.
Stress, Control, Performance Analysis

Wm. P. Kopka, Chelsea, Iowa
Wood Picture Trays—Your Airplane Illustrated in Wood

Charlie Lasher, 190 E. 45th St., Hialeah, Fla.
Ceconite

LeAir Products, 4433 Roselle Dr., Dayton 40, Ohio
Spars, Capstrips, Rib Bracing

Machinecraft, Box 308, Troy, Ohio
Tubing, Sheet Stock

John H. Matthews, 9 Davidson Dr., Woodbridge, Ont., Canada
Caricatures, Drawings of Your Aircraft

B. D. Maule Company, Aviation Division, Jackson, Mich.
Fabric Testers, Tailwheels, Designers and Manufacturers of Aviation Equipment

McGraw-Hill Book Company, Inc., 330 W. 42nd St., New York 36, N.Y.
Aviation Books

MacWhyte, 2906 14th Ave., Kenosha, Wis.
Aircraft Cable, Terminals, Streamline Wire

Midwest Parachute, Novi, Mich.
New, Used Surplus Parachutes

Monarch Mold and Engineering Co., 2321 N. 28th St., Milwaukee 10, Wis.
Wheels

Al Munch, 33 Colony Dr. E, West Orange, N.J.
  Old Mags

Nagel Aircraft Sales, Compton and Torrance Airports, Calif.
  Wheels, Brakes, Tires, Parts

National Insurance Underwriters, 8030 Forsyth Blvd., St. Louis, Mo.
  Aircraft Insurance for Amateur Built and Factory Aircraft

Kal Nelson Aviation, Inc., 8124 San Fernando Rd., Sun Valley, Calif.
  Engines

North-Aire Aircraft Associates, 4079 N. 62nd St., Milwaukee 16, Wis.
  Helicopter Design Book

W. J. G. Ord-Hume, Lake, Isle of Wight, England
  Booklet, Beginners Guide to Homebuilding

C. V. Parker, 2019 B St., San Diego, Calif.
  Wheels

Phoenix Aircraft Ltd., Cranleigh, Surrey, England
  Aircraft Materials, Instruments, etc.

Pioneer Copter Sales & Service, 3303 Parkside Ave., Rockford, Ill.
  Copter Parts, Kits, Plans, Components (Bensen)

Plane-Mobile, 14273-M Beaver, San Fernando, Calif.
  Charts, Drawings for Roadable Aircraft

Plane Pix, 26 Marriot Rd., Barnet Herts, England
  Photo Prints

Popular Flying Association of England, Londonderry House, 19, Park Lane, London W. 1, England
  Books

John A. Powell, 4118 N. Greenview, Irving, Tex.
  Spruce Cap Strips

Precision Machine Works, Box 121, Santa Susana, Calif.
  Machining, Heat Treating

George Rattray, Rt. 3, Afton Rd., Beloit, Wis.
  Fiberglas Cowls, Pants, Spinners, Canopies for "Loving's Love"

Alfred H. Rosenhan, 830 E. 6400 So., Midvale, Utah
  Wheels, Brakes

Sari Publishing, 5617 Hollywood Blvd., Hollywood 28, Calif.
  Design Sketchbook

Arlo Schroeder, 114 S.W. 6th, Newton, Kans.
  Reel Shoulder Harness Control Assembly

Sitka Spruce Lumber & Manufacturng Co., 50 Kansas Ave., Kansas City, Kans.
  Spruce Spar Stock, Plywood, Rib Stock, Miscellaneous Supplies

Skycrafters, Inc., 1365 Gladys Ave., Long Beach 4, Calif.
  Aircraft Radio

Squaircraft, 2030 6th Ave. N.W., Calgary, Alberta, Canada
  Plywood, Spruce, Glue, Engines, Propellers

Stevens Industries, New Goshen, Ind.
  Fabric, Dope, Tires, Belts, etc.

Stits Aircraft, Box 3084, Riverside, Calif.
  Spinners, Tubing, Wood, Engines, Parts, Instruments

Stolp Aircraft, P.O. Box 461, Municipal Airport, Corona, Calif.
  Parts, Engines, Instruments, Tubing, Wood, Wheels, Overhaul

Stoltenberg Company, Elkhart Lake, Wis.
   Custom Machined Parts

Superior Aircraft Co., 5673 Selmaraine Dr., Culver City, Calif.
   Aircraft Plywood

E. E. Svoboda, 3143 N.W. 12th St., Oklahoma City, Okla.
   Navigation Computers, Platters, Study Guides, Log Books

Robert J. Taylor, 621 Denham Rd., Rockville, Md.
   Custom metal fittings for "Fly Baby"

Thorp Engineering Co., 354 E. Cypress, Burbank, Calif.
   Propeller Shaft Extensions

Burrell Tibbs, Box 4337, Oklahoma City 9, Okla.
   Rare, Out-of-Print Books

Trans-Oceanic Trading Co., 711 United Fruit Bldg., New Orleans 12, La.
   Aircraft Plywood, Finnish Birch, Glue

A. Tutka, Box 124, Belleville, N.J.
   Pitot Tubes

Sam Urshan, Box 4332, N.P. Station, San Diego 4, Calif.
   Homebuilt Handbooks

Vancraft, 7246 N. Mohawk, Portland 3, Oreg.
   Blades for Gyrocopters

Victor Sailplanes Co., 213 26th St., Sacramento 16, Calif.
   Plexiglass Canopies

Westair, P.O. Box 23, Hawthorne, Calif.
   Shock Cord Rings

A. Wheels, P.O. Box 174, Ambler, Pa.
   Wheels, Brake Assembly

Guy Whitaker Company, 2952 Crenshaw Blvd., Los Angeles, Calif.
   Aircraft Hardware

ENGINES — PROPELLERS

Aerial Blight Control, Inc., Box 224, West Bend, Wis.
   Engine Overhaul, Tubing, Wood, Parts

Anderson Propeller Co., Inc., RFD No. 1, Box 542, AP 1, West Chicago, Ill.
   Propeller Sales Service, Aircraft, Airboat and Sled, Experimental, Run-In, Gyrocopter Blades

Richard Arnos, R. R. 2, Bryan, Ohio
   Engines, Parts

Banks-Maxwell Propeller Co., Box 3301A, Ft. Worth, Tex.
   Aircraft, Airboat and Snowplane Engines

B & F Aircraft & Supply, 6141 W. 96th St., Oak Lawn, Ill.
   Engines, Propellers, Instruments, Tubing, Plywood

Walter Bullock, Lakeville, Minn.
   Cougar or Tailwind Propellers, Wood or Metal

Carlson Propellers, Box 3072, San Bernardino, Calif.
   Propellers

D. F. DeLong, 619 E. 8th Ave., Eugene, Oreg.
   Drone Engines and Parts

Effingham Flying Service, Effingham, Ill.
    Engines, Parts, Overhaul
Henry Elfrink Automotive, P.O. Box 20715, Los Angeles 6, Calif.
    Volkswagen Engine Conversion
C. Lasher, 190 E. 45th St., Hialeah, Fla.
    Propellers
Harvey Mace, 7560 Henrietta, Sacramento, Calif
    Engines
Miller, Box 912, Oklahoma City, Okla.
    Engines, Parts, Custom Built Propellers
Nagel Aircraft Sales, Torrance Airport, Torrance, Calif.
    Propellers
Oklahoma Propeller Co., Box 457, Shawnee, Okla.
    Airboat and Snow Sleigh Propellers
Pollmann Engines, Wiedenbruck/Westf., Germany
    40 hp Pollmann "Hepu" Engine
Rosen's Electrical Equipment Co., P. O. Box 189, 5028 Van Norman Rd., Montebello, Calif.
    Engines
Phil Schneck, 1460 Highland Ct., Ontario, Calif.
    Volkswagen Conversion Plans, Castings
South Norfolk Air Service, Box 34-A, Rt. 3, Norfolk 6, Va.
    Engine Overhaul
Chris Stoltzfus Agency, Coatesville, Pa.
    Used Engines
Trefethen's, 2432 Chapman St., Lomita, Calif.
    Engines, Nose Cowlings
J. F. Welton, Franklin, Pa.
    Major Engine Parts, Propellers
Wolff-Davidson Company, 11401 O'Donnell Dr., Houston 22, Tex.
    Volkswagen Engine Conversion and Parts, Pollmann Engines, Instruments